NATURAL PRINCIPLES OF LAND USE.

NATURAL PRINCIPLES

OF LAND USE »»» »»» »»» »»» »»»

EDWARD H. GRAHAM

GREENWOOD PRESS, PUBLISHERS
NEW YORK

HD
111
.G67
1969

TO MARY·

PREFACE

THIS BOOK is composed upon the premise that biological concepts can contribute importantly to the wise use and management of land. Experience on the part of land operators throughout the United States has lent proof to the practical value of this idea, particularly during the past decade. Although the land manager may not know the biological laws influencing his work, he is quite familiar with the operations of those laws. He may be likened to the carpenter who is unable to prove the Pythagorean theorem but employs the principle to lay out a right angle by use of the 3-4-5 rule; or the ship's officer who navigates admirably but has forgotten or never knew spherical trigonometry. To treat significant natural phenomena that can serve to provide guiding rules of thumb for the land manager is the purpose of this book. Although the subject has not been treated at length before, it is an important one. Much is now being learned about it and more will be written of it some day, and put to good use. It will be sufficient if this work serves to stimulate further consideration of natural principles of land use and is not too much in error.

The book would have been less adequate for its purpose if the manuscript had not been carefully reviewed by several critical friends. I am first of all indebted to William R. Van Dersal, who helped materially throughout the preparation of the work. To Philip F. Allan, Hugh M. Raup, and Richard M. Bond, I am grateful for many suggested improvements. I wish also to thank H. L. Shantz, E. A. Norton, J. S. Barnes, and Frank C. Edminster, all of whom commented helpfully on portions of the manuscript. Mary Baird Graham bore the burden of typing and much editorial review. I am indebted to the United States Soil Conservation Service for the maps reproduced as Figures 5, 6, and 7, and all of the photographic illustrations except Plate 2 (top), which is from my own files, Plates 8 (bottom) and 30 (bottom), kindly furnished by the United States Fish and Wildlife Service, Plate 22, from the Laboratory of Anthropology, Santa Fe, New Mexico, and Plates 15, 16, 17, and 18 (top), generously made available by Dr. A. C. Cline of the Harvard Forest, Petersham, Massachusetts.

A work like this cannot be done without considerable reference to the work of others, some of whom I have quoted. Several publishers have granted permission to reproduce passages from copyrighted sources which are fully cited in the bibliography. I am grateful to Curtis Brown for a quotation from a newspaper article by H. G. Wells, to Ginn and Company for material from James's *An Outline of Geography*, and to the Yale University Press for two paragraphs from Heske's *German Forestry*. Permission was freely given by the University of North Carolina Press to use a passage from Allport's *Institutional Behavior*, by Charles Scribner's Sons for the table on page 58 from Leopold's *Game Management*, and by the Princeton University Press for a quotation from R. S. Lynd's *Knowledge for What?*. The Philadelphia Society for Promoting Agriculture accorded the right to quote from early publications of the Society, and I am indebted to G. P. Putnam's Sons for material from Sir Arthur Hort's work on Theophrastus. To two English firms, Faber and Faber, and Methuen and Company, I am grateful for approval to quote respectively from Sir George Stapledon's *The Land Now and Tomorrow* and Forde's *Habitat, Economy, and Society*. I have taken a table of types of land use from Van Dersal's *The American Land*, published by Oxford University Press, and wish to express appreciation to Wellington Brink for permission to borrow from my article *Ecology and Land Use*, which appeared some time ago in *Soil Conservation* magazine. The *Scientific Monthly* and the American Wildlife Institute permitted use of other material from my own articles which they have already published. Finally, I wish. to pay homage to the scientific writers whose investigations are chronicled in obscure technical journals. Without their efforts a book like this would be impossible, and the scientific papers I have found useful compose the bulk of the bibliographic references.

E. H. G.

Chevy Chase, Md.
January 1944

CONTENTS

CONTENTS

CONTENTS

NATURAL PRINCIPLES OF LAND USE

INTRODUCTION: A FRESH HORIZON

WE live in an environment of many facets related not as single pieces, but as a mosaic, the pattern of which is not easily discerned at first glance. It must be seen in different lights before we appreciate its true design and real worth. Thus it is easy to look to immediate gain, forgetful of the long-time advantage. But to achieve a lasting economy, man must consider all the effects of his operations on the land. Might it not have been possible to prevent the passage of homestead laws that once encouraged families to live on land incapable of supporting them? Might not the Great Plains have been spared the devastation that resulted from wholesale plowing of the sod with no regard for consequences? Could drainage that exposed land worth less than the cost of drainage operations, and many other activities we now regret, have been avoided?

Is what we plan today equally unwise, or are we prepared to consider all the interrelationships our actions may involve, and act upon that knowledge? It is not an easy responsibility, for it is human to think of today's welfare and of this season's crops. We think somewhat, also, of our own futures, but we have not, as a people, been accustomed to thinking of the future welfare of our country. We have acquired this attitude largely because our short history as a nation has found us busy fighting the wilderness and exploiting its resources with little time for thought of future needs.

To think wisely of the future use of land, we must first look carefully at its past, for a knowledge of what has caused a landscape helps materially in judging its future. Once we understand a landscape's history, we are better prepared to consider how wise or unwise has been the use to which that land was put. Then, with a knowledge of climate, soils, vegetation, and other habitat factors, we can gauge something of the potential productivity of the area.

It takes but a smattering of geological knowledge to know that 'the rocks and rills, the woods and templed hills' are not immutable; that the woods of the East, like the deserts of the Southwest, were once different from what they are now, just as the climates of those areas were once different. In eastern North America, marks of the glaciers are scratched upon the rocks for all to see; in the Southwest, previous forests are revealed by their fossils exposed in rocky outcrops. Less remote change is evident and of even greater significance to the land manager. The white-pine woods of Massachusetts tell a land-use history since colonial times irrefutable in its ecological meaning, and the steep-walled arroyos of Arizona indicate a striking change since early settlers rode easily across those same streambeds in buckboards.

The deep gullies so conspicuous in the Appalachian Piedmont show all too plainly misuse of land, while the condition of farms in southeastern Pennsylvania, after two centuries of constant use, indicate as plainly the advantages of sound management. We must realize that in America we have no more land frontiers, that we must do with what we have, and that the continued welfare of this generation, and every succeeding one, depends first upon how wisely we shall use the land that is ours. To a surprising extent, the present population of the world makes it necessary to plan land use in every continent, for yields of all products of the land depend in large measure upon appropriate management methods.

USE OF THE LAND

Only a few years ago in the mountains of New Guinea, an American museum expedition discovered a primitive people still practicing stone-age agriculture. Oddly enough, these wild people possess a knowledge of some of the fundamentals of land use that are basic to the best management recommendations of the present

[4]

day. Although they still use adzes and axes of stone and employ only sticks for cultivation and tillage, these New Guineans practice bush fallowing, a system of cultivation in which crops are planted in fields cut and burned out of the forest, or bush. After the fields have been worked for several years, new crop areas are cleared in the near-by forest and the old fields are abandoned to revert to trees. Contour terraces are constructed as an erosion-control measure, and vegetation is turned under for green manure. In the bottom lands, the New Guineans have developed an elaborate system of ditching that serves not only to drain soggy areas but, more importantly, to make available the rich black swamp deposits and virgin alluvium of subsurface layers so that they may be turned over upon the exhausted surface soil (Brass 1941). Thus their land has been managed for a very long time.

African tribes have long used intensive crop rotations, as shown by the practices employed in northwestern Tanganyika (Sturdy 1939). There the Wakaras grow a kind of millet (*Pennisetum* sp.). When the millet is about a foot high, a green-manure crop of a leguminous plant, usually a species of *Crotalaria* or *Tephrosia*, is sown in the field. During the dry season, after the millet is harvested and when the green-manure crop is turned under, the field is planted to sorghum. Before the sorghum is cut and harvested, another legume, the Congo goober (*Voandzeia subterranea*), is sown in the field. The goober develops and matures as a pure crop, after which the complicated rotation is again started by the planting of more millet. This simultaneous rotation of two legumes following two grass crops is practiced by a people with the extremely high population density of 600 per square mile— a people who, from their own long experience with the land and without contact with advanced civilizations, have learned to use not only crop rotations and green manuring, but also stall feeding of cattle, terracing, and the construction of stone drains.

So it would appear that the Greeks, who stressed the value of beans and other leguminous crops for 'reinvigorating' the soil, and the Romans, who emphasized the systematic use of crop rotations,

[5]

only extended the application of basic principles of land use long practiced by man in various parts of the world.

In the Americas, also, many measures to use the land wisely were employed long ago. Terracing is an ancient practice, most famously developed by the Incas in Peru. Efficient systems of carefully using water were widely employed. In the southwestern United States, archeologists find ditches that once served to irrigate tilled fields, as well as remnants of contour ridges of rock for slowing runoff. The traveler in Mexico today sees the poorest peon planting crops in furrows that spread water carefully in his tiny fields, a lesson in elementary agricultural engineering inherited from a remote past.

In the highlands of Guatemala, ancestral home of the Mayas, land is still collectively owned and administered by village authorities. Corn remains the chief crop, with beans a very poor second. Tomatoes, pepinos, avocados, and other fruits are also grown. Crops are planted in temporary clearings, according to a bush fallowing or milpa system like that of the New Guineans, Africans, and others in both the Old and New World tropics. Corn and beans are planted together in hills or mounds, and these are arranged in contour rows on steep slopes to prevent the soil from washing. Cornstalks are laid across the slope for the same purpose (McBride and McBride 1942).

Many agricultural practices—systems of irrigation and drainage, terracing, fertilizing, plant selection and breeding, green manuring, and contour cultivation—are so fundamental that they were independently developed in both the Old and New Worlds. It is all the more remarkable that methods of land management were so universal when there was not a single cultivated food plant, or domesticated animal except the dog, common to the two hemispheres before Europeans reached the Americas (Merrill 1938).

If land has been managed for so long, why, then, so much attention today to land management? Primarily because an increased human population necessitates careful use not only of tilled acres, but of those lands best adapted to pasture, range, forests, and wildlife, as well as recreational, industrial, and urban areas. Where primitive society looked to the tillage of isolated patches of land, we now face the time when a large portion of the world's acreage is likely to be put to some determined use. Machine farming has accelerated exploitation of the soil and we have transplanted farming systems from one part of the world to another with little thought of how well those systems were adapted to the new conditions.

With the passing of *laissez-faire* and the transition from individual control to governmental assistance and group use, which daily seem to assume greater importance, a change in management is inevitable—a change necessitating a reoriented social attitude toward the land and the disposition of its products. With the socio-economic aspects of land use, however, this book is only incidentally concerned. The emphasis here is on the biological aspects of the use of land, which accompany the changes being realized in its management—changes now wrought at a rate sometimes surprising even to those who are engaged in their accomplishment.

One of the major problems of the modern land manager is the determination of the best use for each parcel of land according to its potential, long-time productiveness. C. M. Donald (1941) recently brought this point into sharp focus by reference to the British Isles when he wrote:

. . . non-agricultural activities have a valid claim for land surface, but unfortunately each of them is by preference developed on level land and hence usually on good agricultural land. The fertile Middlesex Market Gardening Plain to the west of London is already half consumed by housing and industry. Stapledon [1935] stresses that in the long run (and here he refers to periods of 100 to 500 years) it pays to build reservoirs, suburbs and aerodromes on land of little or no agricultural value—even though the initial cost be tenfold.

In those countries that are well settled, with little public domain left for homesteading, as our own is now, thoughtful management

[7]

is a necessity in order to maintain desired production and provide properly for domestic well-being and foreign trade. The results of scientific research and new techniques must be applied to the land if we expect it either to sustain or increase yields under intensive use. Furthermore, modern land management involves a consideration of all factors influencing the use of land—technical, social, and economic. In the respect that land management is concerned with the relation between man and the land upon which he depends, it deals with natural principles and is a branch of ecology.

It has long been recognized, of course, that biological implications are involved in use of the soil. Maintenance of a fauna of soil micro-organisms helps to support the earth in good tilth and so aids man in the production of high crop yields. Likewise the application of fertilizers is essentially biological, revolving around the use of nutrients by plants. All husbandry is, indeed, a phenomenon of applied biology—in the ecological sense—for it has to do with the relation of plants and animals to their environment, even though the plants and animals are domesticated and the environment, in part at least, is controlled by man.

But there are many relations of living things to their homes that have received slight attention from those who use the land for crops, livestock, or timber. Many of them were actually unknown until recently. A knowledge of them can aid man immeasurably in his efforts to use land wisely and the land manager must be aware of the natural aspects of the work he undertakes.

Ecology and Land Use

Reference to a standard dictionary will show that ecology is that phase of biology which deals with the mutual relations between living things and their environment. Ernst Heinrich Haeckel, the famous German philosopher-biologist, is credited with the invention in 1869 of the word ecology, which he and others for some time after him spelled 'oecology,' since the first part of the word came from the Greek root *oikos*, meaning home.

Haeckel considered ecology to deal with the general economy of the household of Nature. Several years before Haeckel introduced the term, however, a French naturalist, Geoffroy St. Hilaire, used a similar word, ethology, for the same concept, but the word never came into general use, perhaps because St. Hilaire died before he expanded his ethological ideas. Plant ecologists usually look back to the Dane, E. Warming, and his 1895 treatise on ecological plant geography as their primary source, and since the turn of the present century many students of many lands have written on ecological subjects, sometimes using new names to designate their thinking.

It is of more than passing interest for our purpose to remind ourselves that even early interest in this field was not purely academic. In the same year that Warming's plant ecology appeared, the eminent American entomologist, S. A. Forbes, whose studies of insects included so many practical considerations, had already pointed out that economic entomology is simply applied ecology, and his work supported his contention.

Like many words, however, ecology means one thing to one person, something else to another. To the person interested in woodland birds, it may mean study of the food and nesting sites available in the forest; to one interested in prairie plants, it may mean the relation of soil and climate to prairie vegetation. Certainly, as usually defined, ecology is a very broad subject. Even though a biologist studies but a single group of animals—termites, let us say—he may need to employ a knowledge of paleontology, climate, soils, physiography, vegetation, and other animals in order to learn much about the lives of termites and the places in which they live. Furthermore, he may wish to know something of the individual termite's reaction to environmental factors, in which case he will want to delve into morphology, physiology, cytology, and genetics.

Just as one ecologist studies so many details about a special group of organisms, another may essay to study wildlife—all the vertebrate animals of an area—with respect to general habitat conditions such as food, cover, and water. This latitude of interest among students of ecology is shown by the titles of articles in five recent volumes of the American journal *Ecology*, which range from the specific, such as 'The effects of various relative humidities on the life processes of the southern cowpea weevil,' to the general, as 'Prairie soil as a medium of tree growth.' A droll title which combines the specific in regard to animal with the general in regard to environment is illustrated by an article from a recent number of London's *Journal of Hygiene*, headed 'The ecology of the bed-bug *Cimex lectularius* L. in Great Britain.'

Many excellent ecological articles, as a matter of fact, do not appear in ecological magazines, nor do they bear titles indicative of their ecological significance. Such discussions are found in geographical journals, monographic studies of plants and animals, biological surveys, and the writings of Thoreau, Sharp, Beebe, Peattie, and other naturalists who seldom or never used the term 'ecology.'

All of this may be confusing enough, but to make things worse, it is not the diversity of subject matter alone which leads many biologists to refuse ecology a recognized place among the biological sciences. Perhaps because of the mystery of life itself and the complexity of relationships among living things, ecologists all too frequently have blinded themselves to simple realities, confining their thinking to preconceived ideas and broad generalities. Furthermore, in this field of endeavor there has grown up a terminology as difficult and distracting as the cant of any ancient and mysterious sect. In an exact science, the invention of terms is prerequisite to facile manipulation of its concepts, but ecology is, by its very nature, far from exact, in the sense that an abstract science is exact. Strange terms have been invented for ideas already expressed by existing English words. These terms are the millstone of ecology. They drown many an honest effort to use ecological data, and result more than anything else in giving ecology a bad name.

Nevertheless, in spite of the breadth of interests it embraces, the conventional routine it employs, and the multitude of ambiguous and unnecessary terms it affects, there is something very good about ecology. In a way it is the natural history of Theophrastus and Pliny, Humboldt and Buffon, Darwin and Agassiz. But it is also something more, for it has now the advantage of dynamic concepts not recognized, or at least not elaborated on, by those masters. It has access to a wealth of data unknown before the present century, and it has now accumulated a considerable sum of knowledge in its own name. More importantly, ecology represents a way of thinking, for it deals primarily with interrelationships among plants, animals, and environment. It is as much a knowledge of relationships as it is a knowledge of the things related. In the words of the English plant ecologist, A. G. Tansley (1939), 'Ecology is not so much a special branch of biology—in the sense that genetics or the physiology of nutrition are special branches—as a way of regarding animal and plant life.'

From what has been learned of things ecological, there have emerged a number of principles which compose the major contributions of the subject. These are of considerable practical significance, although none has been too well applied to date. Concepts of the community, succession, and predation, and the ideas of ecotones, indicators, and food chains are some of the practical ecological notions as yet poorly recognized. Peculiarly enough, their application is receiving considerable attention in a pursuit even newer than ecology itself, namely the field of modern land management.

The management of land rests fundamentally upon the fact that the capacity of land to produce varies not only from region to region, but within small areas, in accordance with soil conditions, vegetation, slope, exposure, degree of accelerated erosion, and other physical characteristics. Consequently, with respect to physical conditions, and in a very general way, it may be said that farm land comparatively level may best be devoted to corn, cotton, and cultivated crops; gentle slopes to pasture; steep slopes to trees; and odd corners, and infertile and eroded spots to the production of useful wild plants and animals. This may seem so logical that it scarcely needs stating. Would it not be folly to confine trees to level, fertile, unerosive bottomlands while corn is grown on steep slopes where soil washes easily? Yet that has been the land-use pattern in many parts of the eastern United States.

Aside from the fact that some lands must be dedicated to urban, industrial, recreational, and scientific-study areas, the management of land involves, above everything else, the production of a crop, whether cultivated, pasture, woodland, or wildlife. With due regard to prevailing economic conditions, the use of any parcel of land should result in the yield of that crop which can be most profitably supported without permanent injury to the physical capacity of that land to produce. This in itself is an ecological or natural concept, for it embodies the idea that plants and animals are well adapted to the home that supports them, in this case domesticated species in habitats influenced by man. Numerous biological principles are applicable to the management of cultivated crops, to pasture and range management, to the successful handling of forest areas, both large and small, and to the production of useful wild plants and animals.

Naturalists sometimes lament the fact that there are in the world today few sizable areas where plants and animals exist in a condition undisturbed by man. But even before the advent of man, living things were disturbed, so much so that many of them disappeared from the face of the earth. The giant club mosses of the Coal Age and the great dinosaurs that followed them have never been known to man except by their fossils in the rocks. Actually there had been large losses in the plant and animal world long before man appeared. Of the 132 families into which the mammals of the world have been classified, 46 families, or about one-third, are no longer represented by a single living species (Shantz 1941a). It is not strange, therefore, that since the appearance of man, species have continued to disappear and natural balances have been further set atilt.

Civilized man, however, is undoubtedly the most powerful

biological influence for pronounced change that has ever existed (Plate 1). Before his time, any extensive changes in the fauna and flora must have been slow to occur, and were probably caused by broad environmental influences, such as gradual changes in climate. Since the appearance of man, changes have been more rapid—within historic time, astonishingly so—the result primarily of man's influence. By early selecting and fostering plants useful in producing food, man started a change in world vegetation so extensive that in many lands the modern air traveler is seldom out of sight of cultivated fields. If tilled earth is not within his view, his eye is likely to scan land grazed by domestic livestock, for selection and breeding have resulted in the artificial evolution of grazing animals, now so widespread that even the forests and deserts usually support their scattered flocks and herds.

Thus we have encouraged to an almost fantastic degree a few short-lived annual crops, some perennial food-producing plants, and a dozen or so kinds of useful animals. We have depended upon them so long that the origins of almost all of them are lost in antiquity. Although medicinal and fiber plants and wool-producing animals were also given some attention, non-food-producing species for the most part have been left to chance. We have looked largely to wild plants and animals for sources of building materials and other general essentials obtained from living things. Some plants are now recognized as special pasture plants, but even today broad-scale grazing depends primarily upon native grasses, as lumbering depends upon unimproved trees.

The discovery of the New World led to an exchange of cultivated plants and domesticated animals that had been independently developed in the two hemispheres, but it was not until the modern period that man extended his uses of plants and animals beyond those with which he had long been associated. Now we use a great many species for sources of commodities needed in a complex civilization. But this widespread interest is recent. Because of man's long preoccupation with a few improved plants and domesticated animals, and our cursory use of a comparatively few unimproved forms, it is not surprising that the thousands upon thousands of kinds of wild plants and the myriad sorts of wild animals for which no practical use has been divulged are very largely ignored—unless they become pests. Most of them have, in fact, been pushed toward the fringes of used lands, and some have been lost in the shuffle.

Regardless of how well planned and intensively managed our future world will be, countless plants, insects, birds, mammals, and other living things will incidentally accompany man's operations on the land. Some day, in every country, planned programs may determine the pattern of land use, and then not only crops and tame animals, but indirectly the occurrence of all living things will be determined by man's conscious planning and use of the land. It is conceivable that, at least so far as the larger land forms are concerned, the majority of wild plants and animals may even cease to be, except as living museum pieces in refuges or wilderness areas.

Although it may seem presumptuous to think that some day only those animals and plants are likely to exist that man permits to remain, certainly many of those that survive must depend upon habitats maintained as man desires them. But those species that remain—and there will be many—are likely to present problems in proportion to the intensity with which the land is used. As man develops his new world, therefore, it is imperative that he conduct his operations upon the land with acute attention to their effect upon the living organisms that will inevitably accompany those operations.

THE LAND-MANAGEMENT BIOLOGIST

Much has been written about the damage caused by harmful insects, plant diseases, noxious plants, and injurious mammals. On the other side of the ledger have been listed the beneficial effects upon the soil of earthworms and burrowing rodents, the value of birds as natural controls of injurious insects, the recreational values of hunting and fishing, the food we obtain from the great natural fisheries, and economic returns from wild furbearers. These good and ill effects of plants and animals have been recognized with only incidental reference to land operations, and they may or may not be significant to those who operate the land.

It is the particular responsibility of a new technician—the land-management biologist—to look to the relation between the management of rural land, whether it be cropland, pasture, range, woodland, or wildlife land, and the complex of plants and animals which attends such management. The research directed by this biologist, or applied ecologist, if you choose to call him that, and the results he applies are related directly to methods of land management, and in this he differs from other biologists. His task is (1) to increase populations of species that are esteemed for their economic, recreational, aesthetic, or other values, (2) to decrease populations of those species that are harmful to useful plants and animals or otherwise injurious, and (3) to maintain a reasonable balance between communities of living things and land-use practices.

It should never be forgotten, however, that it is not the land-management technician, but the land operator, who ultimately determines what will be done on the land. Furthermore, the task of the land-management biologist is not yet well defined, and his responsibilities are far from routine. His work, as his obligation, is difficult, not only because he must understand and manipulate living things whose interactions are complex, but because this manipulation must be integrated with the planned use of land, in itself an adjustment as yet none too well regulated. Both of these endeavors, although based upon the sciences, are in themselves arts, and their fusion will be an intricate one.

In discussing the ecological role of the introduced cheat grass (*Bromus tectorum*) in western United States, Aldo Leopold (1941) expresses this complication when he concludes that

the cheat problem reminds me, again, how difficult a task has been laid upon the coming generation of technical men. How to join the life of a local community without 'going native' intellectually; how to muster courage to unravel land-use problems which are, at best, only partly soluble; how to become expert in one small technical field without losing the common touch with land as a whole; how to translate technical knowledge into land-use practice without loading the whole job on the government; these are tasks indeed!

Yet they are not impossible tasks, and they must be done. I venture to state three basic implements with which the land-management biologist must be equipped to contribute satisfactorily to their completion. They are (1) a broad knowledge of ecological principles; (2) an intimate acquaintance with the common plants and animals; and (3) a practical understanding of land-management methods. Each complements the other, and none is alone successful. Other facilities supplement their usefulness, particularly an insight into the social structure of human communities as well as into economic influences that perpetuate land-use patterns, and a knowledge of related sciences, such as soil science and climatology. On the basis of present experience, however, the three listed seem to be fundamental. An acquaintance with plants and animals is best gained in the field, as is a working knowledge of land-use practices. To point out something of the relation and importance of natural principles to land-management methods is the purpose of this book.

DISTRIBUTION OF PLANTS AND ANIMALS

ONE of the truly remarkable facts about this world is that enormous numbers and innumerable kinds of plants and animals, large and small, occur almost everywhere. There is scarcely any land—and practically no water—on the face of the earth which does not have its particular population of living things. The kinds of life and their density are widely diverse. From the broad expanse of the mid-Pacific to the cold stretches of Siberia and the Antarctic, living things are making an effort to live, and in most places are wonderfully successful. The distribution of plants and animals has been given considerable attention by naturalists, and during the past century the subject served as one of the most productive phases of biology. The works of Humboldt, Darwin, Wallace, and Schimper are biological classics dealing with the occurrence of plants and animals and the reasons for their distribution.

That kangaroos are Australian, that giraffes and gorillas are African, and that yaks live in central Asia are matters of common knowledge. Other facts of continental plant and animal distribution are accepted without question—it is 'natural' that they are so. But the reasons for distributional patterns are not easily stated or entirely understood.

The presence of living things seems to be dependent first of all upon their geologic history or the history of the land they occupy. For example, none of the Carnivora are native to Australia because that island continent was separated from the Asiatic mainland before carnivores developed—in fact, before any mammals, except marsupials such as the duckbill platypus and kangaroo, had evolved. Because of this geographical isolation, it was physically impossible, prior to the arrival of man himself, for Australia to become populated by the higher mammals that evolved on the mainland.

As geological changes occur, there are frequently changes in climate and life. In the desert canyons of southern Arizona, the relict occurrence of sycamores, walnuts, and other deciduous trees characteristic of eastern forests suggests a once-continuous distribution of these trees across the southern United States possible only under more widespread humid conditions. The rocks lend emphasis to this assumption, for fossils of hardwoods are found in many parts of the West now unsuitable to them. Today the broad-leafed trees persist only in the larger shaded canyons where physiographic conditions maintain a humid micro-climate. Analysis of plant pollen deposited in peat bogs throughout eastern North America during post-glacial time shows definite fluctuations in forest composition that are presumably due to changes in climate.

In addition to their geologic past and the influence of physiography, the history and inherent characteristics of living things themselves help to determine the places they occupy. Some plants and animals seem to be sufficiently flexible that they can, over

---

PLATE 1

TOP. The irrigated valley of Las Palomas Creek, New Mexico, in the midst of shrub-covered desert, illustrates man's influence upon the land. Here various phases of management are hazardous, the first problem being careful distribution of water. Excess soil alkali can easily accumulate as a result of improper irrigation and in such areas weed, insect, and rodent damage to crops is intensified.

BOTTOM. South of Albuquerque, the pueblo of Isleta stands as an isolated example of an ancient culture only little influenced by great changes near by. The intensively cultivated fields of the white man, his railroads, his works and way of life have not materially altered the Indian community, which contrasts sharply with the changed environment surrounding it.

the years, adapt themselves to gradual changes in environment, even resulting in forms especially suited to particular conditions. Others, by genetic change, may give rise to new forms at first restricted to comparatively small areas. Experimental studies are now teaching us a great deal about evolution, and field studies in cytology and genetics are adding zest to the erstwhile static taxonomy of the laboratory.

Geologic changes and evolutionary history, however, are distributional factors of secondary significance to the applied ecologist. Factors of more immediate concern are existing climate, physiography, soils, and the biota.

CLIMATE

Living things respond to climate and cannot thrive where the climate is unfavorable (Plate 2). The redwoods of the West Coast have persisted only where heavy fogs drift in from the Pacific. Giant cacti live only in the desert, where, among other factors, freezing temperatures last no longer than 19 hours. The royal palm ventures no farther north than the tip of Florida where a tropical climate prevails. Other southern plants will not grow in northern latitudes because they are not 'hardy' there. Game managers have learned, often at great cost, that climate limits distribution of animals also. The stocking of exotic birds and mammals holds little promise of success unless, above all else, a species is adapted to the climatic conditions under which it is introduced. Native animals emphasize this lesson. The bobwhite quail, on the edge of its range in Wisconsin, spreads northwestward during years of normal weather, but a severe winter kills so

---

PLATE 2

Seasonal differences are often as marked as climatic variations between regions. The willows planted to control bank erosion along this Vermont river (top) must not only be adapted to generally prevailing climatic conditions but must be able to withstand the rush of flood waters and the terrific physical punishment of grinding ice to which they are annually subjected (bottom).

[19]

many of them that their front then falls back to its normal position.

We do not know the full reason for the effect of climate upon life, and the complexity of the question may be illustrated by the influence of length of day upon plants. Some plants, like red clover, radish, and primrose, are long-day plants, and if they are grown where the days are short they may never flower or fruit. Other plants only fruit if the days are short, for example, tobacco, dahlia, and cosmos. Cosmos plants may grow to a height of 15 feet yet never flower if they are under long-day illumination. Strawberries produce abundant fruit when day length is 14 or 15 hours, but usually do not even flower when the day is only four hours shorter.

Florists for some time have made practical use of this tendency among plants, for they delay the blossoming of chrysanthemums by providing an hour of light during the night. A close relative of the chrysanthemum, pyrethrum, remains as a squat rosette of leaves during short days; it grows stemmy and produces flowers (the source of an insecticide) only when the days are long. Varieties of a species may react differently. Some kinds of soybeans produce abundant seed only when the day is short. It is amazing that when a leaf from an Agate soybean, a long-day form, is grafted onto a plant of the Biloxi variety, a short-day plant, the latter changes and blooms as a long-day plant.

Some plants, of course, can reproduce reasonably well by vegetative means, and perhaps there are many which could set seed with little regard to variation in day length. Those that are known to respond to photoperiod (length of daylight), however, indicate how a climatic factor, the effects of which were long unsuspected, may control plant distribution. Some such plants of temperate climates, for instance, could not perpetuate themselves in the tropics, where climatic conditions by usual standards are 'less severe.' Although they might develop a vigorous vegetative growth, they could never produce seeds there because the days during the growing season are shorter than in temperate latitudes.

Climate has been considered so important in plant and animal distribution that the term 'climatic climax' is sometimes used to indicate the highest type of vegetation and associated animal life which a given area can support—provided man's influence be negligible. Although it is a concept that has evoked much theoretical discussion, it serves to emphasize the fact that life and climate are so inextricably associated that their relationship cannot be ignored, even in matters of extremely practical ecology.

One of America's foremost climatologists, C. W. Thornthwaite (1942), has stressed the importance to the practical ecologist of detailed climatic data. Agricultural specialists who have tried to employ ordinary climatologic observations have found that these do not, except on rare occasions, provide answers to questions that arise in relation to agricultural production. For example, entomologists have sought the climatic factors that control the distribution and spread of the European corn borer in the United States, because the correct delimitation of the climatic range of the insect would permit the relaxation of a costly quarantine. Climate and weather are believed to be the principal factors responsible for the distribution and spread of the insect, but no significant relationship has been discovered. Familiar observations, such as daily and monthly totals of rainfall and daily and monthly means of temperature have little to do with the occurrence of the corn borer. The important features of the environment are the temperature and moisture regimes of the atmosphere where the eggs, larvae, and the mature insects are found.

Likewise agronomists have been searching for the connection between climate and the yield of corn. Here, too, the values of temperature and precipitation that are regularly measured do not, for the most part, define the critical climatic elements for corn yield. They do not provide observations of temperature and moisture of the root zone in the soil or air at the ends of the tassels. In short, we need more knowledge of the micro-climate, based upon careful studies of conditions immediately surrounding the plants we wish to encourage, including light as well as temperature and moisture, if the ecologist is to be properly informed about the relations between the elements of climate and crop yields. Much the same reasoning applies to animal species.

Climates of the world have been classified, and anyone interested in climate and land will want to familiarize himself with the classical and world-wide system of the German climatologist Köppen (1931). He will also wish to become acquainted with the classification by Thornthwaite (1931), whose system, developed largely for North America, is being given widespread use by modern students. The merits of each of these classifications are soundly supported by their respective advocates. Recently there has been an attempt to revise Köppen's classification for North America and a new map for this continent has been prepared accordingly (Ackerman 1941).

PHYSIOGRAPHY

Second only to climate, physiographic conditions strongly affect the distribution of living things. The vegetation that grows in a ravine or canyon is different from that of the prevailing uplands or river floodplains. The soil that develops on valley slopes is different from that of more level land, and the climate, influenced by the topography, is also distinctive. So much different is it that the micro-climate of such a physiographically delimited area is now considered to be a prime factor with respect to plant and animal distribution. Even the two sides of a canyon are different, especially if they face north and south, where direct insolation becomes an influence (Plate 3). Although this effect is especially obvious in arid country, it is nonetheless real in humid regions. A similar effect is noticeable on the two sides of a hill or mountain range, with soil, vegetation, and animal life markedly different on a north-facing slope from that on the opposite slope facing toward the south. A community of living things dependent upon topographic features is sometimes called a 'physiographic climax.' Deep canyons, mountain ranges, escarpments, broad

[22]

rivers, and comparable landscape features act as barriers to some species of plants and animals; under suitable conditions they may serve as well to extend the ranges of others. The Colorado River, for example, separates the ranges of related mammals. The Harris ground squirrel occurs to the east of the river, while a closely allied species, differing in size, color, and habits, is found only to the west. East of the river in Arizona lives a pocket gopher akin to, but distinct from, another gopher that occurs only on the California side. The Amazon River of Brazil is a classical example of a geographic barrier, for even some kinds of birds have the limits of their ranges at the river's edge. The desert stretches of Arizona so thoroughly isolate the desert mountain ranges that there has evolved on each range its own distinctive race of squirrel.

The Green River of Utah illustrates the way in which a watercourse may extend the range of a species, for it has served as a highway on which plants from southern Utah have moved northward through the Book Cliffs escarpment, which the river bisects. Only by that means has the single-leaf ash been able to gain access to the region north of the Book Cliffs. In eastern United States, the red birch follows the valley of the Susquehanna River and its branches into central Pennsylvania. The occurrence of trees of humid eastern forests in grassland types of vegetation far to the west is made possible by conditions along the main watercourses.

Physiographic avenues and barriers are more important in the distribution of living things than are physiographic regions. For example, after a careful attempt to relate the occurrence of the mammals of Ohio to broad physiographic features of that state, Enders (1930) concluded that 'the use of physiographic provinces is not adequate for the statement of the distribution of mammals in Ohio.' Other instances of this sort remind us, as we discovered in regard to climate, that it is not so much general environmental influences that control the distribution of plants and animals. Within the major environmental areas represented by broad physiographic provinces, specific elements of topography, slope, and so on are frequently the principal influence upon the occurrence of living things.

The physiographic provinces of the United States have been neatly mapped by Fenneman (1931), and a graphic relief map in considerable detail has recently been prepared by Raisz as an insert to Atwood's book on North American physiography (1940).

SOILS

In a general sense, the climate, vegetation, soil, and animal life of a natural area are closely interrelated. As already pointed out, however, factors such as physiography may alter vegetation within broad natural regions. Underlying rock formation or parent material may affect the soil type and, in turn, the plants and animals occurring in an area. An outstanding example is the highly fertile and productive Black Prairie of central Texas—a strip of dark, heavy soil developed upon a calcareous substratum geologically different from the lighter soils to the west and east. On this soil, tall grass prairie originally existed within a region where the predominant vegetation was an oak woodland known as the cross timbers. Farther east, in a narrow arc across central Alabama, a similar stretch of prairie—the Alabama black belt—extended through deciduous woodland. On such prairie areas, the soils— rendzinas—are also different from the podzols of the adjacent forests. So it is possible for grassland to occur in areas climatically suited to deciduous woodland. A vegetation type so determined is often referred to as an 'edaphic [soil] climax.'

It should be remembered that soil itself is not an inert medium, but that it represents a segment of the habitat more complicated than the air above it. Bacteria, minute plants and tiny animals, fungi, molds, insects, plant roots, and vertebrate animals live in the soil. As many as 60,000,000 bacteria are contained in a single particle of surface loam, and the mycelial threads of many mushrooms and molds, the roots of grasses, herbs, shrubs, and trees pervade the earth's veneer. Investigations have shown that the earth in England harbors 50,000 earthworms to the acre, an acre of Maryland meadow supports 13,500,000 invertebrates at no greater depth than 'a bird can easily scratch,' and the soils of the Russian steppe maintain 415,000 ant nests per square kilometer. Ninety-five per cent of all the insects invade the soil at some period in their lives.

Where there are no earthworms, burrowing rodents take their place, and California pocket gophers do as much work in soil mixing in five months as earthworms can do in five years. Animals not only live in the soil, they cultivate it, fertilize it, and change it as much as life in the soil has changed them, for they are marvelously adapted to the earth. The occurrence of animals may be directly related to the kind of soil. Some burrowing mammals, such as pocket gophers, are most abundant in light, sandy soils, but absent from heavy, clay soils near by.

More attention is given to the soil in a subsequent chapter, for it is more easily injured or improved than most other environmental factors affecting living things. The soils of the United States have been mapped in the *Atlas of American Agriculture* (1936) and more recently in the *United States Department of Agriculture Yearbook* for 1938.

BIOTA

Living things are as much influenced by each other as they are by climate, physiography, and soil. Vegetation very largely determines the kinds of animals inhabiting a particular area. There are animals as characteristic of tropical jungle, tall-grass prairie, cactus desert, arctic tundra, and oak-hickory forest as are the plants of the areas themselves. The protective shelter, nest-building materials, and food available in vegetative cover are possibly more important in the occurrence of animal species than the climate and other habitat factors.

Conversely, animals may exert a profound influence upon

plants. Some plants owe their dispersal to animals that carry their seeds (Plate 3). Others could not be pollinated or produce fruit without the assistance of insects, many of them highly specialized and intricately adapted for the fertilization of particular flowers. Animals affect not only individual plants, but the composition of vegetation as well. Some ants remove all vegetation from large areas around their mounds. Cattle grazing may maintain as grass-land an area once forested, as in the koa (Acacia koa) forest lands of Hawaii (Baldwin and Fagerlund 1943). It has been claimed that the vegetation of the short-grass plains east of the Rocky Mountains was determined by the constant grazing of buffalo, and that it has been perpetuated, since their virtual extinction, by herds of domestic cattle. Ecologists suppose that, without such grazing, these plains would be more like the mid-grass prairie existing farther to the east in a region of somewhat higher rainfall (Larson 1940). A type of vegetation determined largely by animal action is termed a 'biotic climax.' (Plate 3, p. 34.)

Many of the land-management measures cited later in this book hinge on the reaction of plants to animals or of animals to plants, and thus exemplify biotic influences. When individual plants or animals affect other individuals, of the same or different species, they may begin to compete with each other for existence. Competition is a phenomenon long recognized, and is something the land manager must keep more or less constantly in mind. Theophrastus, long before the Christian era, wrote that 'trees may destroy one another, by robbing them of nourishment and hindering them in other ways. Again an overgrowth of ivy is dangerous, and so is tree-medick, for this destroys almost anything. But halimon is more potent even than this, for it destroys tree-medick.'

In many ways we know little more about competition among plants than Theophrastus did. Why do grasses grow so well with legumes, thus furnishing a combination that makes good pasture? Is it because the legumes furnish the soil with nitrogen used by the grasses, and the grass roots in turn protect the legumes against physical injury by preventing the soil from heaving? Perhaps, but it has been learned recently that some strains of bluegrass completely inhibit the growth of white clover, a legume, despite the latter's nitrogen-fixing bacterial root nodules. A botanist has lately learned that in California deserts some annual plants are most vigorous when growing in association with shrubs, while others occur only in open places away from the shrubs (Went 1942). This would seem to indicate that the shrubs stimulate some annuals and inhibit others, but there may be more to the story that we do not know.

There are certainly intimate relationships among living things which cannot be associated directly with degrees of competition for moisture, sunlight, and nutrients. Many plants and animals live together because they do not come into direct competition. Thus the tap-rooted shrub and fibrous-rooted grass draw upon different sources of water and may live in close proximity. For lack of a more enlightened knowledge of this relationship, we can think only of some measure of 'compatibility.'

While competition and compatibility have had a long time to accomplish some kind of equilibrium in little disturbed communities, such as existed in America before Cortez, Pizarro, and the Pilgrims, man is today, more than ever before, the arch competitor and most important biotic factor. He chooses plants and animals he wishes to preserve and use; the others must show a high degree of adaptability to conditions as modified by him in order to survive. Some, like Russian thistle, ragweed, the housefly, coyote, black and brown rats, and the house mouse, do very well, but many species have already succumbed as man became dominant.

Even in primitive times, man probably kept vegetation types from reaching a stage they might have assumed without his interference. The Indians of the Southeast regularly burned the sprouting and seedling hardwoods that would have changed the composition of the pine woodlands if they had not been checked. In the Southwest as well, Indian fires are believed to have influenced the native vegetation. Later white men's cattle subdued the grasses which we now believe to have been a dominant part of the climax vegetation of the arid parts of Arizona, New Mexico, and western Texas. Throughout the United States, fire, lumbering, grazing, mowing, plowing, reclamation, and the building of cities by modern man have so changed the landscape that it is now impossible to find even remnants of many types of virgin vegetation to preserve as ecological checks against the programs of land management we wish to conduct.

GROWTH FORMS

Biologists have not only tried to determine the effects of habitat factors upon plants and animals; they have tried to interpret the environment by means of the forms the living things assume. Climbing deer mice of woodlands have long tails to help them keep their balance in precarious places; their relatives of the grasslands that live on the ground have short tails. The short, diminutive legs as well as the arrangement of ears, eyes, and nostrils of the hippopotamus prove this mammal's adaptation to an essentially floating life in an environment largely aquatic.

Even among closely related species there are differences of form that correlate with climate. There seems to be a tendency for animals of cold regions to conserve heat by the reduction of radiating surface, for they have smaller ears and tails and generally more compact form. Foxes of the arctic tundra, forest, and desert show progressively larger ears as they inhabit progressively warmer regions. The principle exemplified by this relationship is often referred to as Allen's Rule (Allen 1877).

Among plants, spininess seems to be associated with dry, warm temperatures, as in desert plants. Gorse grown experimentally in dry conditions is much more spiny and has fewer leaves than it bears in its humid home. Great forests flourish well in moist, warm latitudes, while in cold regions of short growing seasons, as in the arctic and at high altitudes, trees do not occur. Neither do they occur in regions of long growing seasons and little moisture, as in deserts, where highly specialized plants grow in bizarre forms.

[28]

Those who study plants have played most with the idea of classification according to form, and a Danish botanist, C. Raunkiaer, has done a great deal of work on the forms of plants. He considered the forms assumed by plant species to express the climate under which they live and their adaptation to survive drought. Raunkiaer (1934) characterized plants by the amount and kind of protection afforded to the buds and tips of the aerial shoots. By this means, trees and shrubs are placed in a group in which the surviving buds or branch-tips perennially project into the air; another group includes plants with buds situated on or near the soil surface, as most of the perennial grasses; and still another group embraces those with bulb-like buds buried in the ground at a distance from the surface that varies with the species. Annual plants of the summer or favorable season form a separate group, as do the stem-succulents like cacti and the tree-supported air plants or epiphytes of the tropical and mountain rain forests. Although the idea of growth forms of plants has not been put to much use, it may be of some significance as a practical guide. As pointed out below, one should not expect plants of tree-growth form to succeed, without artificially modified conditions, in an area which naturally supports only herbaceous perennials of bunch-grass form.

It is of considerable interest that modern naturalists are looking more to growth forms of plants and habitat niches than to general vegetation types or life zones to explain animal distribution. Horned larks, races of which exist from the highlands of Colombia to the Arctic, may be found under highly divergent climates and life zones, but never where they are not associated with broad expanses of short grass.

In a recent informative paper, the bird artist, Roger Tory Peterson (1942) points out that the Parula warbler of the North lives where the pendant moss-like *Usnea* lichen hangs from trees of the cool, coniferous forest. The bird's close relative of the South, found in a deciduous forest under quite different climatic conditions, also shows its dependence upon a specific habitat element, that is, a plant-growth form. The latter depends for nesting materials upon *Tillandsia* or Spanish moss, a plant not even closely related botanically to the *Usnea* lichen, but remarkably like it in its pendant habit and general physical properties. As a caution against dogmatic conclusions, however, we must recognize that the Parula warbler does nest where neither *Usnea* nor *Tillandsia* occur, as another student (Petrides 1942) reminds us. In New York and Maryland it makes its nest in flood-lodged leaves and debris, or of grass, leaf skeletons, cord, or wool.

Growth forms of plants—grasses, shrubs, small deciduous trees— are often expressive of successional stages in the re-establishment of widely different types of vegetation. The distribution of an animal may be directly related to a stage in plant succession dominated by a particular growth form. Pitelka (1941) has shown, for example, that the Redstart, a vivacious, tiny warbler, prefers a habitat dominated by second-growth trees from 20 to 40 years old. Thus in the northern coniferous forest, the Redstart is found only in open stands of birch and aspen, while to the south in deciduous forest, under quite different climatic conditions, it nests primarily in cut-over areas where aspen, cherry, and comparable second-growth trees are prevalent.

LIFE ZONES

Men must always have wondered why particular plants grew only in particular places, and why a certain animal could usually be found in a special sort of habitation. Even before much was known of the world, people realized that plants and animals are not scattered promiscuously over the face of the earth but are grouped together in orderly fashion. Early writers recognized this fact, and Theophrastus, full three centuries before Christ, wrote (Hort 1916) as follows:

That each tree seeks an appropriate position and climate is plain from the fact that some districts bear some trees but not others; (the latter do not grow there of their own accord, nor can they easily be made to grow), and that, even if they obtain a hold, they do not bear fruit—as was said of the date-palm, the sycamore and others; for there are many trees which in many places either do not grow at all, or, if they do, do not thrive nor bear fruit, but are in general of inferior quality.

Alexander von Humboldt, however, long after Theophrastus, is conceded to have been the first to classify natural areas of the world in his essay on the geography of plants in 1805. Since then many others have divided the globe into natural faunal or floral zones, regions, or realms.

Among the first in the United States to bring order to our ideas of the distribution of life was the late C. Hart Merriam, who, in 1890, while engaged in a biological survey of the San Francisco Mountains and adjacent desert of the Little Colorado River in northern Arizona, was struck by the distinct altitudinal zones of vegetation on the mountain slopes. Many others had treated groups of plants or animals in North America with respect to natural areas, but the classification which Merriam devised was the first major attempt to use climatic data and to base provinces on both plants and animals. In the United States it became the most widely used concept of its sort.

Certain animals and plants were listed by Merriam as characteristic of each of the zones he delimited, and the system proved so appropriate, especially in the western states, that it was employed by naturalists of the United States Government in all of the faunal surveys that were undertaken during the period when the Far West was opened up by exploration and settlement. Throughout the biological surveys published in the now classical series, The North American Fauna, the use of Merriam's Life Zones gives uniformity and coherence to the classification of broad, natural communities, from Alaska to New Mexico and Alabama.

Merriam gave some attention to the practicality of his scheme, and related crop plants to his life zones. He also attempted to expand his classification of altitudinal zones, so vividly displayed in the western mountains, to the entire continent. The idea was far less useful in the East, however, where physiographic features are less extreme. Although Merriam at first recognized that life zones were best expressed by the plants and animals characteristic of them, and continued to identify zones by indicator species, he finally tried to delimit the zones by temperature alone. But temperature is a much more reliable exponent of climate in areas which differ radically in altitude than in areas where latitude and continental influences determine the prevailing climate.

The names of Merriam's Life Zones, in order of increasing altitude or latitude, are: (1) Tropical, (2) Lower Austral (Lower Sonoran in the West, Carolinian in the East), (3) Upper Austral (Upper Sonoran in the West, Alleghenian in the East), (4) Transition (frequently subdivided into Dry and Humid), (5) Canadian, (6) Hudsonian, and (7) Arctic-alpine. The northern, or altitudinally highest, boundary of a zone Merriam determined by the sum of the mean daily temperatures above the plant zero during the growing season. The plant zero is the temperature above which plant growth begins, which Merriam arbitrarily chose as 43° F. The southern, or altitudinally lowest, boundary of a zone he determined by the mean temperature for the six hottest consecutive weeks (Merriam, 1892, 1894, 1898).

Merriam's work has been severely criticized because (a) the zones cut across natural areas, such as the central grassland, which extends north and south, (b) the temperature threshold or plant zero is not the same for all plants, and (c) of admitted errors in Merriam's calculations. Furthermore, one cannot ignore the important influence of other climatic factors, such as moisture, upon plants. One can scarcely observe a verdant, irrigated valley in the desert without the conviction that it is water alone which determines the existence of plants, although, of course, no environmental factor functions to the exclusion of others. It is sometimes said that temperature is the climatic factor most importantly influencing the distribution of animals, and that water is of most significance with respect to plants. No one who has attempted to weigh all the influences of climate, however, has come forth with a concept of life zones that has been more widely used than Merriam's. Thus, in spite of its limitations, we must admit the fact that a poor tool, which was at hand, was better than a good one which was not available.

Many naturalists have tried to delimit zones on the basis of the natural occurrence of living things. This has been done particularly by the botanists, who have mapped vegetation. Thus we have maps of the United States showing natural types of forest, grassland, and desert that are very helpful in visualizing our country before its transformation by the Spanish and other European immigrants. One of the earliest vegetation maps is that by Shreve, published in the *Geographical Review* in 1917. Another is that drawn by Shantz and Zon for the *Atlas of American Agriculture*, which appeared in 1936. Many distributional maps have also appeared in state and regional floras, and in faunal and floristic studies for many local areas.

At first glance there may seem to be no practical significance to the mapping of natural areas or the plotting of major plant and animal communities that no longer exist except in very much disturbed condition. An understanding of the original vegetation of an area and the animals associated with it, however, frequently forms a basis for management recommendations and establishes a focal point for relating many biological principles, as shown in later chapters. It may well be said that the life that an area can support not only expresses climate, soil, and other environmental conditions, but indicates adapted crops and pertinent land management recommendations.

Even in a world so changed by man as ours has been, land-use operations can profitably relate to the pattern of vegetation which originally clothed the earth, and it is frequently to man's advantage to keep this relationship in mind. In the central United States, in the broad, transition area between the eastern deciduous forest and the grassland to the west, we know that uplands of tall grass prairie once alternated with wooded bottomlands and tree-covered floodplains. Throughout this region, the establishment of trees for shelterbelts, windbreaks, and farm-post lots has been widely attempted. Some of the trees have been planted on sites that naturally had never supported trees, but were originally occupied by grasses. With a supply of supplementary moisture, often provided by carefully engineered and expensive diversion ditches gathering rainfall from broad slopes, together with cultivation to reduce competition from other plants, the trees were made to survive. On the other hand, when planted on sites where soil, slope, and other local conditions originally supported woody vegetation, trees have grown well at comparatively little cost and with a minimum of attention. Consideration of environmental conditions as summarized and expressed by natural vegetation types or areas, even though such areas persist only as tattered remnants of a once complete pattern, is of real assistance in planting operations, whether for windbreaks, range rehabilitation, reforestation, or other practical purpose.

A usable variation of the life-zone idea is the establishment by horticulturists of growth or 'hardiness' zones, which indicate the areas in which it is safe to use certain plant species. Most nursery catalogues carry a map of such zones to guide the planter. This thought has received most careful expression in the map of Plant Growth Regions developed by F. L. Mulford and associates in the United States Department of Agriculture, and most recently revised in W. R. Van Dersal's book on the *Native Woody Plants of the United States, Their Erosion-control and Wildlife Values*

PLATE 3

TOP. This Pennsylvania field affords a striking example of a biotic factor influencing the distribution of living things. Droppings from birds perched in the dead tree have seeded the surrounding area to red cedar, sumac, arrowwood, dogwood, cherry, sassafras, and other species that provide food for birds. BOTTOM. The influence of physiography upon living communities is strikingly displayed on opposing slopes of this Utah valley. The south-facing slope supports a forest of juniper trees while the opposite side is a solid chaparral of little sagebrush. Normally the sage occurs at somewhat higher, the juniper at lower, altitudes.

(1938). It divides the United States into 32 zones or areas, each having fairly uniform growing conditions for plants. Originally prepared as a rose-zone map to indicate areas in which particular types of roses could be successfully grown, it was later used as a basis for recommending the planting of street trees. As it now stands, it is an attempt to delimit natural areas sufficiently uniform that plants which grow in one part of an area will grow in every other part of that area. Unlike maps of natural vegetation, the Plant Growth Region map does not show zones of like natural vegetation, but zones of like growing conditions. Though useful only in a very general sense, it points the way toward an interpretation of environmental conditions, not as expressed by original vegetation, but as indicators of potential homes for plants valuable to man.

COMMUNITIES OF LIVING THINGS

From the foregoing discussion, it will be evident that man has a very definite interest in various aspects of plant and animal distribution. Oddly enough, it was not at first supposed that the pattern of life upon the earth changed very much except, perhaps, as man influenced it. We now know, however, that associations of living things do change, and furthermore that they possess a dy-

---

PLATE 4

TOP. Surface mining exposes subsoil, which is immediately invaded by pioneer plants. Disturbed some 20 years ago, the weathered shale of this Kentucky area supports a good cover of herbs, shrubs, and black locust. Young sycamore, elm, cherry, and other tree species are beginning to develop a woodland. Water characteristically covers the last overturned strip against the undisturbed land to the right. Unless leveled when mined and converted to pasture, such areas are best devoted to the production of fish and wildlife, and eventually, to wood products.

BOTTOM. An old-field stand of short-leaf pine, nearly a century old, indicates a once-cultivated area near Athens, Georgia. Protected from fire and grazing for many years, an understory of young oaks, hickories, and other hardwoods is succeeding the pines, many of which have already been removed for lumber.

namic behavior peculiar to the group to which they belong, which we call the community. That a community of plants and animals has a character of its own and the capacity to act as an entity contains an idea unrecognized by the earlier students, and the subject is more fully treated in the following chapter. For the moment let us add to what we have mentioned of plant distribution, growth forms, and life zones, the concept of the community.

The life zone is like a big city, rather stationary and uniform, representing broad influences such as climate, just as the city represents transportation, trade, and industry. But the great city is composed of many small racial and cultural communities, separate yet closely interrelated, each changing, growing, and vital with a dynamic energy apart from the city as a whole. Like the human community, the plant and animal community expresses the lives and interactions of the individuals in it. It represents the unit of associated living things. To the ecologist, the function of the community, as well as the behavior of the individual plants and animals composing it, serves to provide a basis for management recommendations and judgments.

That plants and animals within a fairly small circumscribed area may be grouped in some orderly manner, and occur chiefly in association with certain other species, is exemplified by the weeds on a vacant lot. We must think of these plants as composing a group, integrated neither less nor more than the persons living together in a human community. Like people, plants and animals may have different backgrounds, and live individual lives, but together they form a group that has a distinctive character and behavior.

Much has been written of recent years about communities of living things, and the community has received many different names. American ecologists have called it the biome, biotic type, or plant-animal complex, contending that two separate communities, one plant and one animal, do not exist in the same area. Furthermore, they treat this complex as a kind of organism, separate from the inorganic environment, and become very metaphysical about it. It is common, for example, to compare the stages in the plant succession of an old field to the growth of an individual plant or animal, and to consider that a community of living things grows up, matures, and becomes senile. It serves our purpose best to consider plants and animals as different, although closely associated, parts of a community, and to include inorganic factors of the environment—climate, physiography, soil—in any consideration of the community as a working concept. Suffice it to say that the community idea is a useful one, and that a great many of the ecological principles applicable to the management of land deal with the relation of communities of living things to the habitat supporting them, as that habitat in one way or another is modified by man.

The term 'natural area' is often encountered and, although it has no exact meaning, it is a useful expression. It is less specific than life zone or community, and is used in a broad sense to denote a region having characteristics of life, climate, terrain, et cetera, which in a general way differentiate it from other regions. But a natural area, whether a great region like the arctic tundra or a small unit like a river floodplain, is difficult to define and not easily explained. We have already noted that it is a complicated problem to correlate types of flora, fauna, climate, soil, physiography, and other environmental factors. That such correlations have been mapped is due more to the fact that the delineator of one factor, as soils, used as a guide a previously prepared map of, let us say, vegetation—or vice versa—than to the fact that such a relation so exactly exists.

Yet connotations such as arid plateau, tall-grass prairie, and muskeg country are expressive and meaningful. Such natural areas frequently serve as a rough definition of land condition and suggest in crude terms the adapted land use. In using the term 'natural area' the land-management biologist should recognize the fact that it cannot be exactly defined, for it is as yet impossible to evaluate all the influences that go to determine the character of a landscape. If, on the other hand, we were able to mark all the

[37]

factors which define a natural area, we should be well on our way toward possessing a real evaluation of environment in its complete sense. When we have obtained such knowledge, we shall possess a new and valuable tool for those engaged in the manipulation of the land and its manifold products.

TRANSITION AREAS

Along with the concepts of life zones and communities goes that of the transition area between two zones or communities of living things. This transition zone, often called an ecotone, may be broad or narrow, indefinite or sharply defined. It may be the sub-arctic forest between the arctic tundra to the north and the true coniferous forest to the south, or it may be the vine-crowded, shrubby 'edge' between a woodlot and a cultivated field. In the latter sense, it has received considerable practical attention on the part of wildlife managers, for it is usually true that animal populations are higher in such an area than in either of the zones or communities that join to form it.

The abundance of wildlife has frequently been stated as an expression of edge, and recent work has shown that 'the population density of most nesting birds varies as a direct function of the amount of edge per unit area' (Beecher 1942). The interspersion of cover types, such as openings in woodlands, tree-covered floodplains in grassland, and, in cultivated areas, hedges, streambank vegetation, scattered vegetated odd spots, and roadside shrubs, means much to wildlife by providing necessary habitat elements. Such areas provide a great deal in the way of nesting cover, food, and protection from predators. Numerous studies confirm this, and we know that the abundance of a species, especially one with limited cruising radius, depends upon the distribution of appropriate types of cover, a relationship sometimes referred to as the *Law of Interspersion* (Leopold 1933).

Before civilization made them less directly dependent upon environment, the numbers of men illustrated the influence of transition areas. In northwestern Pennsylvania, abundance of artifacts

indicate that the Indians lived more around the margin of Pyma-
tuning Swamp, a large, open bog of northern character within a
deciduous forest, than in the swamp itself or in the upland beyond
it. The Indians undoubtedly found good hunting and gathered
fruits there, and may even have cultivated patches of land.

The land manager, for the most part, deals with either disturbed
or transitional environments. Consequently he must be especially
well informed about the characteristics of these everchanging and
comparatively unstable components of the landscape. Many of
them are man-made. George S. Wehrwein (1942) has pointed
out that land problems appear in their most acute form on three
fringes or transition zones. One of these is the transition between
cultivation and grazing, as exemplified by the Dust Bowl. Another
is the familiar cut-over land representing the fringe between farms
and forests. The third is the suburban area between farms and the
built-up city—the rural-urban fringe. The last reminds the land-
management biologist that although his work deals primarily with
rural land, it at times impinges upon problems of urban origin
and city planning.

THE CEASELESS CHANGE   » III

THE preceding chapter dealt primarily with the distribution of living things as they are scattered on the land surface. For the most part, it considered older, static aspects of ecology. Yet living things and the communities into which they are grouped do not remain fixed. Some of the most important ideas the ecologist can offer to those who use and manage land have to do with the meaning of the changes that are forever taking place. To understand those changes and direct management measures in agreement with them involves an appreciation of much that is fundamental in modern ecology.

That changes occur is commonplace observation—as anyone who has tried to keep a garden free of weeds is fully aware—but that those changes are not random ones and fit well-established patterns is less fully realized. The weeds that come into your garden are very much the same kinds that invade the garden of your neighbor; they comprise a comparatively few aggressive species. The series of events which takes place after plants populate a disturbed area, such as an abandoned garden, is discussed below. First of all, however, the plants must reach the garden and conditions for growth must be favorable there.

Much has been written by ecologists about the migration, invasion, establishment, and competition of plants, and something is known of the ways by which plants move into and populate an area. On the other hand, much is not understood about all this. For example, in a part of Java where some 300 kinds of weedy plants occur, only 30 species regularly invade old rice fields or, under experimental conditions, will even germinate in soil from those fields (Kooper 1927). This suggests that the distribution of seeds is a matter of less importance in determining the occurrence of pioneer species than, perhaps, physical or chemical conditions of the soil.

From other studies, however, like one conducted in the Appalachian Piedmont, we learn that the invasion of 20- to 40-year-old stands of loblolly pine by hardwoods is not determined by conditions of the soil, but by the transport of seeds and types of root systems involved (Coile 1940). Birds and small mammals carry many of the hardwood seeds into the pine woods. The shallow roots of loblolly pines compete with each other. The roots of oaks, which invade the pines, are deep with an early taproot. The oaks can obtain a position of dominance under the shade of the conifers, where their roots reach for moisture and nutrients below the shallow root systems of the pines.

Ideas about the establishment of plant communities range all the way from the contention that the repopulation of a bare area is primarily the result of chance (Gleason 1926) to the belief that it is as regular as the growth of an individual organism (Clements 1928). Be that as it may, the student of land-management ecology will find much of significance in the research and observations on this subject and it is to his advantage to be familiar with the plant communities of disturbed areas and their transformation into other communities with the passage of time (Plate 4, p. 35). This change is known as succession.

SUCCESSION

Although the reasons for the invasion and establishment of plants may not be clear, it is an obvious fact that areas where the vegetation has been disturbed or destroyed do become reclothed with plants, usually of a given group of species. Once established, these plants gradually give way to others that invade the area and change the composition of the vegetation. One change blends into another, more slowly than at first, until, if not further disturbed, the area eventually becomes clothed with a type of plant cover similar to that which occurred there before the original disturbance.

The biologist finds practical as well as theoretical usefulness in this concept, based upon the idea that primeval vegetation undisturbed by man, whether it be forest, grassland, or desert, is in essential equilibrium with prevailing environmental conditions. Such vegetation is considered climax for the region and will remain essentially unchanged so long as the climate and other important influences remain relatively stable. If the climax vegetation is removed from an area, it usually is not re-established until several different plant communities have successively occupied the area, each community or stage more like the climax type than the preceding.

If a virgin hemlock-hardwood forest is destroyed by lumbering, it is not a stand of young hemlock and hardwoods that immediately begins to replace the cut-off trees. Instead there first springs up a cover of annuals, usually a few kinds of widespread weeds. Gradually, in the course of a few years, a growth of herbaceous perennials replaces the annuals. Still later, shrubs invade the area and finally, after some years, trees such as cherry, aspen, and birch appear. In the course of a long time, these trees are gradually replaced by climax species of hemlock and hardwood, to produce eventually the type of forest originally on the land.

Of course, many kinds of disturbance, as fire, cultivation, and erosion, may modify this succession, and stump sprouts from some of the hardwood species may vary the picture (Plate 5). Man's activity usually tends to prevent development of the climax. Sometimes there seems to be a 'telescoping,' so that succession does not develop regularly, but is hastened by skipping stages. Under other conditions, early communities persist for a long period, witness the California hills that remain in an annual cover of wild oats and the extensive areas of western range that persist for long periods in cheat grass. But it is a natural rule that the vegetation of a disturbed area tends inexorably to progress toward the type of plant community that is in equilibrium with the prevailing conditions of climate, physiography, soil, and animal life.

A similar progression occurs in a newly formed water area, such as a pond or lake. Of course the species differ. The bands of vegetation around a lake, from the submerged plants in the open water to the vegetation on the dry land, roughly suggest the stages through which a water area progresses toward the climax vegetation of the region. Frequently the first plants to appear along the shallow shore are cat-tails or bulrushes. With these, or often preceding them, are submerged or floating water plants. As the lake fills in by accumulation of silt and debris, the cat-tails move toward the center, to be followed by sedges which, in eastern United States, are succeeded by shrubs such as buttonbush, alder, and dogwoods. Eventually the shrubs, in turn, give way to trees. In a given region, the vegetation which finally develops will be composed of the same climax species, whether the succession began in the bare earth or the water of the pond.

It should not be forgotten also that, as stages develop in the succession of vegetation, new groups of animals—soil micro-organisms, insects, birds, and mammals—associated with the various stages also appear. In fact, animals, large and small, are so closely associated with successional stages that they are an inseparable part of them, and together the plants and animals constitute the community of living things that occupies a given area at a particular time. It may be remarked that, with the change in communities according to successional trend, there is an associated change in soil conditions, frequently reflected in an actual modification of the soil profile.

A classical example of how animal succession parallels the successional development of plant communities is that described by Shelford (1907 p. 9), who studied the carnivorous tiger beetles (Cicindela) on the sand ridges of the south shores of Lake Michigan. He found that distinct species of beetles occurred in the various stages of plant succession as they were represented from the sandy lake margin inland through several vegetation types to mature beech-maple forest. There is scarcely such a thing as animal succession independent of plant succession, at least among the vertebrates. Animals influence the plants by distributing seeds, pollinating flowers, and in other ways, and, conversely, the animals depend upon the plants for food and shelter.

One of the few cases of an independent animal community among the vertebrates occurs on the guano islands off the coast of Peru. It is possible that the first birds to populate these islands were those that could nest on the surface of the plantless, granite rocks. Later, when the excrement of these birds, accumulating in the almost rainless climate, was sufficiently thick to provide nesting sites for burrowing birds, there must have been a change in the bird life.

This is suggested by what has happened in reverse order since men began to mine the guano deposits for use as nitrogenous crop fertilizers. In some parts of the islands, the guano deposits, more than 100 feet deep, were dug off practically to the granite substratum. The penguins, diving petrels, and Inca terns, which could no longer find guano in which to burrow their nests, then moved to other parts of the islands. Surface nesting species, as pelicans and cormorants, became more numerous as the competition lessened. Later, as operations continued on the islands, the pelicans left the cormorants and boobies, more tolerant of man, to become the most common birds (Murphy 1936).

If guano again accumulates as a result of recent conservation measures adopted by the Peruvian Government, a repetition of the supposed original changes may occur. At any rate, here is a rare case in which vegetation does not influence animal succession. Even the food of these birds is indirectly derived from plants, namely marine algae, for the birds live entirely upon sea fishes, chiefly a small anchovy.

Although many people who work on the land today have little inkling of the meaning of succession, it is a concept long recognized by observant people, for there is occasional mention by the ancient writers of regular change in vegetation. The early Americans knew of it, too, and one Isaac Wayne, in a letter to Mr. Peters, then president of the Philadelphia Agricultural Society, in 1814 wrote of the change in forest cover at Valley Forge, Pennsylvania:

That the timber prevalent about Valley Forge, previously to its being fallen for the use of the American army, in the autumn of 1777, and winter and spring of 1778, consisted of white oak, black oak, Spanish oak, rarely interspersed with scrubby chestnut and hickory.

That the ground on which white oak was formerly the most conspicuous, now exhibits black oak, hickory and chestnut, in abundance, and in great perfection.

That where black oak had been most general, white oak, hickory and chestnut now plentifully exist.

That of the hickory and chestnut, there were at least 16 to 1 of what grew on the same ground in the year 1777; and that these two latter species of timber, were now flourishing, in the highest degree, in many places where no other timber formerly grew, than white oak, black oak, and Spanish oak.

By 1850, G. B. Emerson, in his introduction to a comprehensive work on the forest trees and shrubs of Massachusetts, wrote that 'it is abundantly confirmed by my correspondents, that a forest of one kind is frequently succeeded by a spontaneous growth of trees of another kind.' Although Emerson did not recognize any regularity in the process, he nevertheless cited a growth of spruce and hackmatack—American larch—being succeeded by maple, oak being followed by pine, hemlock by white birch, and other sound examples of plant succession. It would seem, therefore, that the ecologists were not the first to recognize succession, nor should they remain the only ones to use a knowledge of the principle.

The idea of succession is one of the ecological principles most useful to the land manager (Plate 6). Without an appreciation of what it means, a second-growth forest, the thorny shrubs of an overgrazed range, or the weeds on an abandoned field are that and nothing more. Recent studies of the vegetation of abandoned fields in various parts of the country indicate that it is possible to determine, from the plants growing there, how many years ago a particular field was last cultivated. Everywhere on old fields one can see plant communities that represent successional stages of vegetation (Drew 1942). With a knowledge of succession, it is possible not only to understand something of the past history of a parcel of land, but to predict its future as well, which, in the last analysis, is one of the greatest tests of the usefulness of any human discipline. (Plates 5 and 6, pp. 50-51.)

By careful management, for instance by controlled burning in a pine woods, regulated grazing of rangeland, or selective cutting of forest, it is possible to maintain vegetation in a successional stage most profitable to man. This is sometimes spoken of as arresting succession. Most of man's cultural activities on the land, in fact, consist of maintaining a stage of plant succession profitable to his purpose. On crop fields, even the earliest pioneer species must be kept from developing; in pastures, the herbaceous perennial stage is favored. In woodlands, a community high in the successional series is required, free from excessive fire, insect depredation, or windfall damage, if the forest is to be most productive. Many examples of arrested succession, although seldom referred to by that name, are cited in later chapters as good land-management measures.

INDICATORS

Old timers in New England will tell you that wherever black walnut trees grow, the land is good, but where poverty grass forms a carpet, the land is poor. No ecologist taught them this. Early Americans had to judge the potential crop productivity of a piece of land by the vegetation upon the land before it was cleared. Yet early settlers were not the first to notice such things. One of America's most practical ecologists, Arthur W. Sampson (1939), reminds us that the Romans and the Greeks before them used plants as indicators. Theophrastus recognized plant species with exacting requirements as indicators of specific site conditions, and Pliny knew that the presence of certain forest types denoted land suitable for the cultivation of wheat. That plants can indicate the proper use to which a parcel of land is best adapted is an idea of great value in land management.

By 1858, E. W. Hilgard, pioneer soil scientist, stated to the governor of Mississippi, in a report calling for continued support of the geological and agricultural survey of the state, that results of the survey would serve as a guide to purchasers of land. 'Had the survey been called into existence earlier,' Hilgard wrote, 'it might have saved some money to those unfortunate speculators who, allured by the prairielike levelness of the Tippah, Pontotoc, and Chickasaw "flatwoods," invested their capital in a kind of stock which, to their amazement, has remained utterly unproductive.' Far to the north, twenty years later, T. C. Chamberlin wrote, in his treatment of the geology of eastern Wisconsin, that 'the most reliable indications of the agricultural capabilities of a district are to be found in its native vegetation.' He held a strictly modern view, namely that vegetation is an expression of the combined 'influences of soil, climate, topography, drainage, and underlying formations and their effect upon it.' He likewise realized that communities as well as plant species were useful indicators, stating, for example, that the heath plants of a bog suggested areas suitable for cranberry culture.

Among the botanists, H. L. Shantz, now directing the wildlife work of the United States Forest Service, was one of the first (1911) to show specific correlations between the occurrence of plant species and the agricultural possibilities those species represented. As the Far West opened to settlement, Shantz, Kearney, Piemeisel, and other ecologists showed crop potentialities for many areas with alkaline soils. Plants and animals, as well as biotic communities, indicate not only crop productivity, but also past condition or use, successional trend, and other things of value to the land manager. Once one learns to recognize in plant communities indications of past or changing conditions, it is surprising how many landscape features take new meaning.

OF PAST CONDITION OR USE

Throughout the entire eastern United States, solid stands of pine now tell of past cultivated fields. In central New England, white pine frequently indicates once cultivated areas; in Virginia, scrub pine tells the story; and in southern Georgia, stands of long-leaf pine indicate a similar history. So certainly do these trees reveal past cultivation that all of them, regardless of species, are locally known as old-field pines (Plate 7). One can wander through pine woodlands 100 years old and prove the accuracy of these plants as indicators of past land use by the old plow furrows and occasional terraces still plainly evident beneath the mulch of needles now covering the ground. (Plate 7, p. 66.)

Plant communities serve as indicators of climax vegetation as well as of past land use. Where white pine occurs on old fields, we can assume that the original vegetation was a mixture of oak, maple, and birch, with scattered conifers such as hemlock and white pine. Where scrub or Virginia pine populates old fields, it is fair to assume that the climax vegetation is a hardwood forest composed predominantly of oak and hickory, with other deciduous species such as tulip tree and ash. Long-leaf pine comes in where, without fire, the virgin forest was very probably composed of southern oaks, red and black gums, wax myrtle, and dogwood. These are broad generalities, but they indicate what, with more exact knowledge, we could expect successional stages of vegetation to tell us of the climax toward which each is pointed.

Raup (1940) has even ventured to map a line across northeastern Connecticut, northern Rhode Island, and eastern Massachusetts which suggests the division between the original central New England hardwood forest of red oak, maple, and white ash to the north and the virgin forest of oak and hickory to the south of the line. The extent of the northern area is indicated by white pine in old fields, the southern by the revegetation of once-cultivated areas with red cedar and gray birch.

In addition to the story told by old-field pines, past history is

[48]

indicated by other plants and plant communities as clearly as by any other means except the memory of man or his written word. In the north woods and western mountains, patches of aspen or birch almost always indicate old burns, just as in the east fireweed and pin cherry tell the same story in burned-over, deciduous forest. Some pines are typical indicators of fire and occupy large areas previously burned. Lodgepole pine (*Pinus murrayana*) and jack pine (*P. banksiana*) even hold their seeds in closed cones for many years, releasing them readily only after fire has affected them.

Ecologists have long claimed that the occurrence of certain plant species indicates overgrazing. Thus wire grasses (*Aristida*) in west-central Texas show a change from the original bluestems (*Andropogon*), and further grazing transforms the cover of wire grass to buffalo and mesquite grasses (*Buchloë* and *Hilaria*). Prickly-pear cactus in great abundance on land once covered by mixed prairie grasses is almost invariably a sign of overgrazing (Plate 7), and in eastern pastures the fields of mullein, yarrow, ironweed, bugloss, or vervain suggest a like misuse. There is considerable reason to believe that continued heavy grazing has permitted mesquite trees and cedars to spread into areas once dominated by native grasses. Attempts to eradicate or control the spread of these woody species should not be made without reference to management, such as the re-establishment of native grasses by regulating livestock numbers using the area.

OF CROP PRODUCTIVITY

In a broad sense, land once forested proves best for growing trees such as peaches, pears, and apples, while the cultivation of the cereal grasses—corn, oats, and wheat—is most profitable in regions once occupied by native grasslands. When the western United States was being opened to settlement, the work of practical biologists showed correlations between the vegetation types of semi-arid areas and possible crop uses. Original stands of sagebrush, for example, indicated a well-drained, porous soil nearly free from alkali, which the homesteader could profitably put into cultivation. Greasewood, on the other hand, was indicative of a fine, compact soil high in alkali, at least below the surface, and with a high water table—a soil unsuited to general crop production unless freed of alkali by irrigation and drainage.

In the East, patches of pine woods scattered throughout a generally predominant deciduous forest denote areas of fine, sandy soil distinct from the heavier loam and clay types characteristic of hardwood areas. Southern farmers long ago learned that when long-leaf pine is cleared for the cultivation of corn, it will be necessary, after three years of cultivation, to fertilize the ground if yields are to be maintained. If an area with a mixed stand of long-leaf and short-leaf pines is cleared, corn can be grown without declining yield for 5 to 7 years. If land once covered with mixed oak and hickory is turned into cultivation, 10 or 12 years of good yield can be expected, while soil formerly supporting short-leaf pine and oaks is good enough to produce corn for 12 to 15 years without the addition of fertilizer.

OF OTHER CONDITIONS

As already mentioned, pine woods may indicate well-drained, sandy soils usually drier than the heavier soils in the same area that support hardwoods. Peculiarly enough, in less humid areas sandy soils may be the ones which hold more moisture than associated, heavier soils. Thus, in western Nebraska, where the prevailing vegetation is mixed-grass prairie, the lighter soils of the sand hills support grasses of the tall prairie characteristic of subhumid conditions farther east. Not only greasewood, but salt

---

PLATE 5

The dynamic quality of plant succession is illustrated by the 3-year change in vegetation in this cut-over Ohio woodlot. In 1935 mature beech, oak, and white ash were removed (top). A year later herbaceous plants covered the forest floor, and in 1938 (bottom) shrubby species and sprouts from the cut stumps were filling the opening.

grass, shadscale, and many other plants indicate alkaline soils. Acid soils are suggested by such plants as poverty grass, blueberries, and sheep sorrel. A few students have even contended that vegetation types reveal underlying geological formations (Cuyler 1931), but this is only likely under rather unusual conditions.

Growth forms of plants are indicative of environmental conditions. Thus forest trees are characteristic of humid climates, bunch grasses of sub-humid conditions, and columnar cacti of desert areas. Specifically, the replacement of fibrous-rooted plants by taprooted species has recently been listed as an indication of accelerated erosion. Animals may also indicate environmental conditions, as already pointed out, by variations of body form and size of appendages under different climates. Likewise there is a general change in tone of mammal pelage from dark in southern areas to a lighter color in northern climes. Even the kind and abundance of mammals may be revealed by the condition of plants if we know the meaning of the signs. Large numbers of deer are indicated by the close cropping of cliff rose and ephedra in the Kaibab Forest of Arizona or, in the forests of the Adirondacks, by the mutilation of hobblebush, another preferred deer food.

It is in the recognition of plant and animal communities and the changing conditions they disclose, however, that indicators are of the greatest use to land managers. We have already mentioned that there is always a natural tendency for vegetation types to vary under the influence of climatic and other environmental factors. Under the influence of man's use, vegetation may undergo comparatively rapid change. If use is severe, the vegetation is likely to change through a series of stages to a community domi-

---

PLATE 6

A knowledge of plant succession is put to good use by the land manager. This destructive Illinois gully (top) in a heavily over-grazed pasture was stabilized by the simple expedient of fencing it from livestock. Two years later (bottom) a cover of soil-protecting vegetation had controlled soil loss and pointed the way toward future productive use of the land.

[51]

nated by annual plants; if the use is light, the area will become populated by perennial species in a kind of balance with the imposed conditions. When grazed, for example, grassland vegetation usually changes from a community of perennial grasses to a stand of grasses mixed with perennial forbs. If heavy grazing continues, a complex of perennial weeds with very few grasses will appear, and, with still more intense use, a sparse growth of annual weeds finally predominates.

When ranges are in the annual weed stage, they are frequently dominated by a few pestiferous exotics, and demands for weed control are numerous. It is no accident that demands for rodent control are also most insistent from those who are running stock on weedy range. Kangaroo rats, ground squirrels, pocket gophers, and some other rodents are known to be much more numerous on badly used grassland than on moderately grazed areas. With a knowledge of plant succession, the land-management biologist is equipped to prevent considerable waste of funds and effort by pointing out that plant cover and animal numbers are directly related to intensity of use, and that an adjustment of use, in this instance reduction of livestock, may not only diminish the weeds and rodents by removing the reason for their occurrence, but result in a more productive range.

ANIMAL CHANGES

The ceaseless change is shown by the response of animals to alterations in environment. In the United States, settlement, clearing, and cultivation have resulted in new landscapes and highly modified conditions, and conversion of the land pattern has been accompanied by altered animal populations. Obviously, the fauna of an agricultural countryside is not the same as that of the forest or prairie which preceded the occupancy of the area by man. The kinds of native animals—insects, birds, mammals—that are found in the altered habitat, however, occurred in some part of the region before settlement. Many animals have found it impossible to adapt themselves to man's activities, and have persisted only in isolated, undisturbed areas, or disappeared. The buffalo, antelope, wolf, bear, moose, caribou, and many other wilderness animals are now far less abundant in the United States than they were at one time, and others, like the passenger pigeon, heath hen, and great auk, have gone forever.

That all wildlife was more plentiful in primitive times, however, is an idea springing more from imagination than from fact. Numerous native animals have become more abundant as a result of settlement, witness many insect and mammal pests, while some useful species find the changed conditions to their liking. A word about increased populations of native animals with respect to changed conditions will illustrate the opportune dependence of living things upon 'unnatural' conditions. It is now well known that the bobwhite quail ordinarily requires woodland, brushland, grassland, or areas of comparable growth form, plus cultivated land, for an optimum range. Wildlife managers try to arrange proper distribution of these four 'cover types,' or their equivalents, in order to support a maximum number of quail. It is easy to understand that the eastern United States, under primeval conditions, did not provide a great deal of this sort of habitat, and it is hard to believe that the bobwhite was a more common bird before American settlement than it is now. Probably it occupied areas of grassy, open woodland, where fires or other natural disturbance encouraged native herbaceous grasses and legumes that provided food in place of cultivated grains.

Some animals have extended their ranges as a result of man's activities. There is good reason to believe that in this country the opossum has spread far to the north since white settlement, and the coyote now occupies a far greater area throughout the West than it once did. The destructive Colorado potato beetle has spread widely from its home in Colorado throughout the United States almost wherever potatoes are grown. The reasons for the spread of some animals are not easily determined, but the distribution of the fox squirrel has been linked with a food plant, the osage orange (*Maclura pomifera*). This tree originally grew in southwestern Arkansas and adjacent Oklahoma and Texas, but after the breaking of the prairies it was found adaptable for windbreak hedges and was widely planted in eastern Oklahoma, Kansas, and Nebraska. The seeds, developed in large, heavy, orange-like fruits, are a favorite food of the fox squirrel, and the planting of osage orange hedges has made it possible for this mammal to spread widely across uplands where formerly the prairie grasses limited it to wooded stream courses. It is quite probable that the western range of the fox squirrel has been extended by the widespread planting of the osage orange (Whitaker 1939).

The development of this country favored many introduced species and some of them have become locally naturalized. In the United States, the house mouse, brown and black rats, English sparrow, starling, and ring-necked pheasant represent vertebrates that find life singularly favorable with man and the conditions he fosters. Many injurious insects from foreign lands have spread widely as a result of man's activities. Among them we find enemies that cost man much in ingenuity, time, and funds to combat. The Japanese beetle, gypsy moth, San José scale, cotton boll weevil, chinch bug, corn borer—all these and many more have increased a thousandfold because of environmental conditions induced by man.

Thus changes in land use are accompanied by changed animal populations as well as by transformations of vegetative pattern. The land manager will do well to look to this matter, for frequently problems arising from an overabundance of an injurious animal species can be most easily settled when the relation between the animal and the conditions of the land are recognized. Likewise, the increase of a desirable species is often accomplished only if the effects of changed conditions upon the welfare of the species are reasonably well understood.

THE preceding chapters have dealt with the distribution of living things and with the changes that are constantly occurring in the composition of communities. Distribution pertains largely to the static aspects of life, while changes reflect dynamic processes. In addition to concepts that are easily related to one or the other of these general considerations, there are numerous other natural principles that prove useful to the land manager. Some of the more significant ones are treated below.

Food Chains and Pyramid of Numbers

One of the most readable books of an ecological nature is Charles Elton's *Animal Ecology*. Written by an outstanding international figure in the study of animal populations, this book has a great deal to say about food chains and the pyramid of numbers. A *food chain* may be defined as a series of species in a community, each of which is related as predator or parasite to the next. A simple food chain is illustrated by the statement that the cat eats the mouse, the mouse eats the grain. Food chains sometimes serve to emphasize an affinity between well-known, but not obviously connected, things. This is exemplified by a diagram drawn by Hall (1942) to show the indirect ecological relationship between the fisher, a beautiful carnivorous mammal, and the yellow pine trees of western forests, thus:

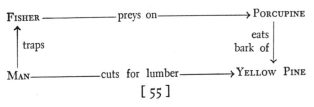

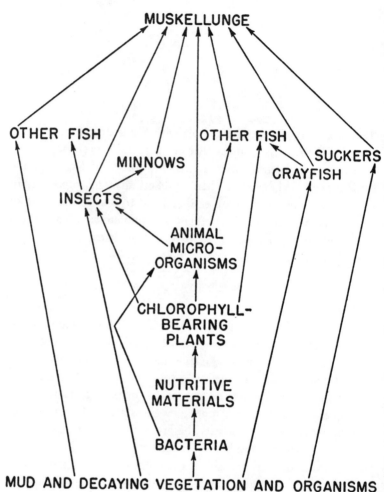

Fig. 1. Food chain for the Muskellunge (after Williamson and O'Donnell, 1941).

So man reduces the fisher, not only by trapping it for its very valuable fur, but by destroying the trees that supply an important item of its food.

Complex food chains are usually presented as diagrams (Fig. 1).

[56]

In a food chain, numerous relationships exist. If, for example, one member of the chain kills and devours another member, the relationship is spoken of as predation. If the members are mutually helpful, we speak of symbiosis, as the alga and fungus living together dependently in lichen form. If one member lives upon another without killing it, the result is parasitism. Simple as this 'house that Jack built' idea may seem to be, a thorough knowledge of food chains is difficult to acquire. It is, nevertheless, prerequisite to the successful management of most forms of wildlife.

There is also much talk by practical ecologists of *buffer* species, which occupy a place between predator and prey. The deer is said to buffer cattle from the mountain lion because the lion has more opportunity to catch deer than to kill cattle. In the diet of the great-horned owl and other animals that eat bobwhite quail, rodents, rabbits, and small birds are buffers with respect to quail because they serve as substitutes for quail in the owl's food. To protect a game species, it may be of considerable advantage to the game manager to understand the role of buffer species. A real knowledge of buffer species hinges upon a thorough understanding of the ecology of an area, and we have much to learn about the part they play in food chains.

It is practically impossible to isolate a food chain, for the animals in it usually have some food relation to species in another chain. In the diagram of the muskellunge, for instance, the 'other fish' listed are also eaten by predatory species other than the muskellunge. But that relationship would be part of another food chain with the other predatory fish at its head. All of the food chains in a community that have elements in common constitute a *food cycle*.

A very interesting thing about a food chain is that the species at the top is in most instances physically the largest, and occurs in least abundance, while the species at the bottom is usually small, frequently microscopic in size, and occurs in great numbers. This idea, which indicates density of occurrence of the species in a community, is referred to as a *pyramid of numbers*, or, sometimes, as an abundance table (Table 1). Because Charles Elton was the first to write at length about such a number relationship, it has also come to be referred to as the Eltonian pyramid. Two food chains simply stated and the pyramids of numbers they suggest are:

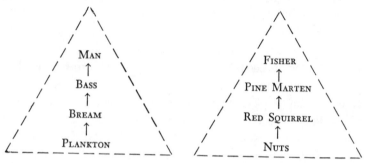

TABLE 1. ABUNDANCE TABLE

Populations of Breeding Animals on 1 Square Mile,
Santa Rita Range Reserve, Arizona
(after Leopold 1933, p. 233)

NUMBER	SPECIES	BASIS
1	Coyote	Rough estimate
2	Horned Owl	" "
2	Redtail Hawk	" "
10	Blacktail Jackrabbit	Strip count
15	Hognosed and Spotted Skunks	Rough estimate
20	Roadrunner	" "
25	Cattle (over 1 year old)	Forest Service permits
25	Scaled Quail	Rough estimate
25	Cottontail	" "
45	Allen's Jackrabbit	Strip count
75	Gambel Quail	Rough estimate
1,280	Kangaroo Rat (*Dipodomys*)	Taylor 1930
6,400	Wood Rat (*Neotoma*)	" "
17,948	Mice, spermophiles, and other rodents..	" "

There are instances in which the top animal is not the largest. A lion may not be larger than a zebra or eland, nor a timber wolf than a caribou. Strength, sagacity, or common effort may compensate for size. Small parasites or disease organisms may lay low large animals. A deep-sea fish, by virtue of a distensible belly, can eat fish larger than itself. But in spite of exceptions, the construction of a pyramid of numbers, even in very general terms with rough estimates, is a useful tool. It can serve to orient management measures, whether intended for the reduction of a troublesome animal or for the increase in numbers of a desired species.

Such a device forces consideration of the food that supports the species to be managed and the enemies that tend to reduce its numbers. Management of furbearers like the skunk, for example, is less uncertain if we know something about the abundance of food elements—such as insects, and the numbers of enemies—such as great horned owls, that are likely to occur in the habitat in which the skunks are to be managed. The abundance of insects that are the principal food of the skunk may depend upon the kinds of plants present, and they, in turn, upon soil and other environmental conditions. To the ecologist, the pyramid of numbers is valuable because it helps him to understand the general structure of a natural community. Without such knowledge, attempts to manipulate any component of the community is more guesswork than management.

Range and Territory

Many ideas about animal movements and spatial relations are of use to the land-management biologist. Granting *habitat* as the place—the sum of physical and biotic elements—in which an animal lives, it is frequently of interest to know how much land this means with respect to an individual animal or breeding pair. Game managers commonly use the term *range* for habitat. *Distribution* refers to the entire geographic area occupied by all members of a given species of animal or plant, and has no particular reference to individuals; it should not be confused with range or cruising radius.

The distance an animal habitually wanders is frequently referred to as the *cruising radius* of the species, and may be daily, seasonal, or yearly. The following table illustrates variations in the distances traveled by some species of non-migratory animals:

	AVERAGE DAILY	AVERAGE YEARLY
Attwater wood rat	50 to 80 feet	?
Bobwhite	⅛ to ¼ mile	½ to 3 miles
Ring-necked pheasant	⅛ to 3 miles	½ to 6 miles
Mule deer	¼ mile	8 to 20 miles
Elk	½ mile	40 to 120 miles

The wanderings of non-migratory animals are less regular and usually not so extensive as those of migratory species. A migratory animal's seasonal cruising radius is frequently quite different from its daily radius during the breeding season. A golden plover in the Arctic may not fly far each day from its nest for food, yet travel 8000 miles every autumn from northern Baffin Land to the pampas of Argentina and back again in the spring.

It is of importance in the management of wildlife to understand the cruising radius of the species to be encouraged. For example, knowing the daily mobility of bobwhites to be not more than one-quarter mile, it becomes an axiom of quail management that suitable food must be available within that distance from the nest or from protective vegetation, and that the number of quail may in large part depend upon the occurrence of suitable cover at quarter-mile intervals.

Closely related to cruising radius is the idea of the *carrying capacity* of land, whether it relates to wild species or to domestic animals such as sheep and cattle. Many factors influence the number of animals an area can maintain. The ancients had this in mind when they remarked that 'one hill will not carry two tigers.' A knowledge of the number of animals a habitat can reasonably be expected to support—its *saturation point*—is useful to the land manager. Within broad limits, this is known for some wild animals, examples of which are:

SEVERAL THOUSAND ACRES SUPPORT	1 California mountain lion
40 ACRES SUPPORT	1 mule deer
12.5 ACRES SUPPORT	1 Virginia deer
1 ACRE SUPPORTS	1 bobwhite quail
1 ACRE SUPPORTS	3 or 4 cottontails

The range manager also uses this idea when he balances the number of cattle against the forage available, and considers the distance stock will travel to water.

Many animals seem to set up an area, usually within a certain distance from a nest or den, which they aggressively defend against other members of their own kind, and often against other species. This area is referred to as the *territory* defended by an animal, ordinarily the male, or by a mated pair of animals during the breeding season (Howard 1920; Nice 1941). If, in providing nest boxes or artificial dens for birds and mammals, something is known of the territories they defend and require, the boxes or dens might be so distributed—neither too close nor too widely separated—as to make best use of the habitat in maintaining an optimum population.

There is much that is not yet understood about an animal's recognition of territory. It has been pointed out by students of game birds, for instance, that territory is more than a mere relation to available food, since few species ever increase up to the limit of their food supply. Errington and Hamerstrom (1936 p. 401) have stated that, in the case of the bobwhite, the birds seem to have an awareness of when their territories become over-populated, either with their own species or with other species such as the ring-necked pheasant. Likewise, apart from direct predation, the chief mechanism by which over-populations are reduced in a given area seems to be that of departure of surplus birds to less-crowded areas.

Another term that is used frequently these days is niche. A *niche* is the particular kind of place occupied by an animal within a habitat. We have seen that the climatologist considers the micro-climate to be of greater significance in the welfare of living things

[61]

than climate in the usual sense of annual averages of temperature, humidity, and so on. So the biologist thinks of the 'micro-
habitat' or niche as the most important environmental influence
in an animal's life. To say that the raccoon inhabits the eastern
deciduous forest of North America suggests the habitat of the
mammal, to be sure, but the raccoon is specifically confined,
within that general habitat, to the immediate environs of streams.
It may be said that the niche occupied by the raccoon is wooded
streams. A comparable niche in the southwestern United States
might be occupied by the ring-tailed cat.

PREDATION

The urge to act upon the obvious rather than to delay action
until we can determine the real reasons for a situation is nowhere
more evident than it is with respect to man's treatment of predatory animals. Because it kills an animal which we wish to protect,
the predator is branded undesirable, and the solution to the complete protection of the desired species is often considered to be
the predator's 'control,' which too frequently means attempted
extermination.

In Arkansas, people will tell you that quail are not plentiful of
recent years because the foxes are killing them, and the foxes
ought to be shot. Farther south, you may be informed that quail
are scarce because the skunks suck the eggs. In central Texas,
the armadillo's appetite for quail's eggs becomes the cause of
scarcity (recent studies show 85 per cent of the armadillo's food to
consist of insects), and farther west the road-runner is blamed
for the quail's plight. In western Texas, a local group was so convinced recently that ravens were the cause of their poor quail
shooting that they attempted to destroy a large roost of white-
necked ravens by dynamiting it. It is very easy for man to blame
everything but himself for his misfortune. Throughout Arkansas
and Texas, the lack of quail is the result largely of inadequate
food and cover. However adequate the predator control and pro-

tection from hunting, these cannot compensate for a factor of the habitat which is lacking or insufficient.

That a species cannot increase beyond the limit set by the *least* abundant necessary factor in its environment is a statement of the principle referred to as *Liebig's Law of the Minimum*. We should not conclude, however, that because this idea has been expressed as a 'law' it is more valuable than other principles that have not been accorded such dignity. First applied to field crops a century ago (Liebig 1859 p. 27), this principle is a highly important one for the land-management ecologist. Much unnecessary effort can be spared if, in tackling an ecological problem involving the welfare of a species—plant or animal—attention is directed toward learning the habitat factor or factors which tend most to limit its abundance, and then to improving conditions upon the basis of that knowledge. Sometimes the limiting factor is as difficult to find as the answer to the initial problem one starts to solve. With the realization that the limiting factor may be a highly elusive one, however, caution can be exercised in undertaking operations until it is determined.

The principle applies to numerous fields. For example, boron is needed by many plants, but in such minute amounts that only highly specialized analyses recently detected its importance. Nevertheless, if boron is absent from a soil, many plants will not grow well, regardless of the abundance of other nutrient elements. Polluted streams may carry enough food and permit sufficient penetration of light to maintain a high fish population, but where the oxygen content is made unusually low by sewage, the stream may support protozoans and some other plankton organisms but no vertebrates. If, along our country's wooded bottomlands, food and protection are provided in even more than sufficient abundance for raccoons, but no suitable den trees are present, raccoons will not be found there. Thus every necessary item of its habitat must be present if a given plant or animal is to survive, even the item which may be the least of them all.

Although much is yet to be learned of predation, it can be accepted that predators usually live upon animals that are surplus to the number necessary for the survival of the prey species. If this were not true, the predator, killing numbers of its prey necessary to survival, would eventually exterminate its food. It is well to remember that attacks by predators upon chickens in a chicken yard or trout in a hatchery pond are quite different from predator-prey relationships under less artificial conditions. These should be looked upon not so much as examples as exaggerations of predation. The same is true when man supports large numbers of domestic livestock where they may serve as prey for wolves.

As a result of several years' study of the bobwhite quail in Wisconsin and Iowa, ecologists have learned some fundamental things about predation. Great horned owls and Cooper's hawks are the most serious predators of bobwhite in this region, the former of more importance. Marsh hawks, foxes, and the house cat are of minor significance. The bobwhites surplus to a given habitat must move out or they die during the winter from one cause or another. The investigators suggest that a population 'does not have to have a high density to have a surplus; it only has to exceed the accommodating capacity of the environment, be that high or low. A population of no more than a bird per square mile may have a surplus, if the carrying capacity of the land is less.'

In their own words, the authors (Errington and Hamerstrom 1936 p. 374) state some interesting conclusions:

Contrary to what would seem most reasonable at first glance, lower populations of the most formidable predatory types, down to scarcity or actual absence on the observational areas, has not resulted in any appreciable lessening of the net pressure of predation upon bob-white winter populations. The 1931-32 season showed terrific horned owl predation upon the occupants of the lethal territory number 14; in 1932-33, the horned owls were gone but the losses continued, this time through the medium of grey foxes; in 1933-34, the foxes were greatly reduced in numbers, but still the bob-white losses were annihilatory because of foxes and general predation. In other instances, predator populations have noticeably increased, sometimes to top-heavy peaks, without any apparent effect on bob-white survival as compared with other winters. Not once have we been able to establish a clear-cut case of differences in predator kinds and numbers having any net influence upon the losses from predation suffered by wintering bob-whites.

The one thing that seems really to count in determining the severity of the predation is the position of the quail population with respect to the carrying capacity of the land. This severity we have found to be quite predictable on areas having carrying capacities demonstrated by recorded survivals over a period of years, apart from complications brought on by starvation emergencies, destruction of habitats by wholesale debrushing, burning, fall plowing, heavy pasturing and the like.

On the basis of an intimate knowledge of the food habits of predatory animals, W. L. McAtee (1934), whose familiarity with practical aspects of wildlife conservation is well recognized, has stated:

. . . the situation with reference to predatory mammals in general seems much the same as we find in birds. Only a few of the species can justly be classed as chiefly injurious, and the others under ordinary conditions do more good than harm. This is inevitable from the very nature of predator-prey relationships. The flesh eaters must subsist either directly or indirectly upon the vegetation consumers. Man's crops are chiefly vegetable, hence these vegetation consumers include most of the creatures regarded as pests. Any predator upon them is in some degree man's ally. When animal crops are involved, predators may become injurious but this is a special case. It should be recognized that in Nature nearly all predacious creatures tend to be beneficial from man's point of view, and control plans and practices should be based on this understanding.

Predation is a phenomenon of concern to those who must manage aquatic as well as terrestrial habitats. In Alaska, for instance, there has been considerable demand by commercial fishermen for control of fish predators. Here, in a region where salmon fishing is the prime source of income, serious predation might jeopardize the entire economy of the Territory. Bounties have been paid on trout and bald eagles, while hair seals, bears, gulls, and terns are also considered destructive. The actual damage to the industry by these predators, however, is still uncertain, and studies are being made to determine the true relationships involved. The elusive 'limiting factor' has not yet been discovered.

Along trout streams in some parts of the West, the spritely little water ouzel, a wren-like bird that lives largely on aquatic insects, only rarely on fish eggs or fry, is persecuted as a fish predator. Kingfishers, herons, and mergansers are other birds sometimes killed because they eat game fish. Kingfishers, however, eat largely coarse, non-game species, and herons are most troublesome at hatcheries. Mergansers have given the fish manager some concern, for they seem to prefer trout, and may have some effect upon the fish population, although, as with other predators, it is difficult to say, without sufficient knowledge of actual conditions, whether the fish taken constitute more than the natural surplus.

CYCLES

Like the swastika, spiral, and other geometric patterns that have long fascinated man, and around which he has built many theories, postulates, and speculations, various rhythms and cycles have received much attention. Many phenomena of cyclic nature have been disclosed—sun spots, lake levels, rodent populations, rainfall, tree rings—and attempts have been made to correlate rhythmic

---

PLATE 7

TOP. Heavy grazing of native grassland in northeastern Colorado has permitted the development of extensive areas of prickly-pear cactus. Although drought intensifies such change, the cactus stands as an indicator of the misuse of land—in this instance, overstocking with range cattle.
BOTTOM. Throughout much of the eastern United States solid stands of pine indicate once cultivated fields. This cover of pines in Webster County, Louisiana, identifies a field abandoned a half-century ago.

phenomena, usually with unconvincing success. Although correlations between different things that show cyclic fluctuation are largely speculative, the occurrence of cycles of particular sorts is well authenticated; those of primary interest to the biologist relate to animal populations (Fig. 2).

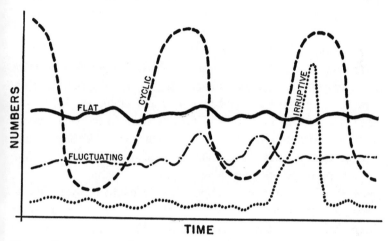

FIG. 2. Curves representing typical fluctuations in animal populations (after Leopold 1933, p. 51).

Populations of animals may vary in regular or irregular sequence, and curves representing typical variations serve to give names to each kind of fluctuation. The great majority of animals are not

---

known to vary in numbers according to regular periodicity, and their populations are said to have flat curves. Some animals, whose numbers usually show flat curves, however, may, like the bobwhite in the Lake States near the periphery of its range, exhibit an occasional abundance. They are then said to show an irruptive population curve. If a population repeats a high at regular intervals of time, it is said to exhibit a cyclic curve; if at infrequent, irregular intervals, a fluctuating one.

The population curves which have aroused the greatest interest are the cyclic curves, best represented by the population behavior of American arctic foxes and lemmings, with high numbers occurring every four years, and by grouse and rabbits, with highs at ten-year intervals. Furbearers of Canada seem to follow a 'grouse' cycle, except muskrats, whose curves are inverse to it. Salmon and cod of the Atlantic Coast follow a similar ten-year fluctuation.

Not too much is known about this phenomenon, although the hordes of Scandinavian lemmings that pour into the sea every so often, and the plagues or outbreaks of voles in Europe have long been the subject of much speculation. The cyclic nature of their abundance, however, was not always appreciated, although rise and fall of mouse populations have been noted by Biblical authors, by Aristotle, and by other early writers. In our own country, from New York to California, mouse outbreaks have been so pronounced that much damage to crops and orchards has periodically occurred. Not only do such rodent peaks recur, but their natural predators—foxes, owls, hawks—tend to vary in numbers with them. Charles Elton's recent book *Voles, Mice and Lemmings* (1942) is our most authoritative work on animal cycles, and he freely admits that we are as yet unaware of their cause.

A knowledge of animal cycles is an important tool of the applied ecologist, for it helps with otherwise baffling problems. It may be financially far better for a state to disapprove expenditures for mouse control than to spend public funds fighting a biological windmill, if it is known that the 'infestation' will decrease natu-

[68]

rally within a very few years. Likewise, as in New York State in 1935, ecologists can warn orchardists and others when to expect rodent outbreaks and to take the necessary precautions to guard their crops against abnormal animal numbers—in this case to protect fruit trees against field mice. Correlating the planting of forest trees at a time when gnawing rodents are at the ebb of a cycle, as with the snowshoe hare in the Lake States (see p. 135), is a wise management measure based upon a knowledge of the fluctuation of animal numbers.

A knowledge of cycles of other sorts may also be of considerable use. To know that rainfall may vary more or less regularly over the years is of such value that it might well avert future Dust Bowls. In retrospect, it seems difficult to believe that favorable 'highs' in the rainfall cycle for the Great Plains so strongly influenced westward movements of people during the second half of the past century. These 'highs,' as well as intervening dry periods or 'lows' of rainfall, occurred every ten or eleven years.

Among the nineteenth-century pioneers, many were so discouraged by the great drought of 1893-5 that they returned disheartened to the East. By 1905-9, when the rains came, and in 1914-16, when they came again, more and more people moved to the West, encouraged unfortunately by the Government as well as the 'colonization' firms and 'land sharks' of that period (Clements 1938). High prices brought by the First World War caused men to ignore periods of drought until 1929, when the economic crash and ensuing dry years taught a lesson which not only those on the land, but many others, should long remember. The dust clouds of 1934 drifting from the Southwest over New York, Washington, and the Atlantic awoke many persons to the fact that something was wrong with the use of the land, even though they may not have realized its full portent.

The past few years have seen a swing into another period of increased rainfall. The Plains have become so unbelievably green that those who have known them only during low-rainfall years, or who have forgotten what has gone before, insist that 'the climate is changing' and we can now plow the earth and load the range as never before. The land manager, remembering that just as surely as we shall have rain for a few years to come, we shall once again face years that are dry, cautions against such short-sighted measures.

ONE of the most informative methods of studying living things is careful observation, with written notes for later reference, and the capacity to observe keenly is worth more than any technique designed to substitute for observation. There are, however, standard aids for evaluating plant and animal populations that demand attention. It will surprise the conventional ecologist that there follows no treatment of means for determining environmental factors—rainfall, temperature, soil, available water, evaporation ratio, pH, light, and so on. Their consideration is omitted here not only because of the brevity of the book, but more importantly, because the living things themselves are of greater significance to the land-management ecologist than the factors which govern their occurrence. We have acknowledged in Chapter II the importance of the chief factors governing distribution, but plants and animals as living expressions of those factors more fully exemplify environment than an artificial summation of influences reached by the most careful mathematical calculation.

In many ways a sound judgment based upon observed facts is far more useful than a set of statistical data meticulously tabulated. Uvarov (1931), the Russian ecologist, in discussing the relation of insects to climate, has pointedly stated the fallacy of expecting great contributions to biology from exact enumerations:

The danger of statistical (and generally of mathematical) methods in ecology is that their application gives a stamp of extreme exactitude and reliability to conclusions even if derived from faulty, though sufficiently numerous, data. Once the method is mastered, the work of drawing conclusions from facts becomes mainly a mechanical process, instead of one of reasoning on biological lines.

It is practically impossible to obtain exact numerical data in quantitative ecological work, and all figures are either approximate, or represent the mean of more or less inexact figures. No mathematical treatment can render the basic figures more exact than they are, so that no conclusions drawn by the application of mathematical methods can be more convincing than those arrived at by a critical discussion of the data. Mathematical methods should be used in ecology, perhaps, more widely than they are today, but it must never be forgotten that they have their limitations and may even be positively misleading.

The fetish we make of figures is amusingly shown by a recent article in a scientific journal which correlated the age at which physicians die with the degree of prominence they attain. The mean age in years at death was stated to the fourth decimal point, with a margin of error carried to the same exactness, more than a thousand cases being involved in the calculations. And, believe it or not, the degree of prominence for each physician was determined by the number of lines in the death column notice of a professional journal.

Acknowledging the limitations of statistical methods, we must recognize practical means of recording composition and density of vegetation and of forming approximate estimates of the numbers of animals. Although this book is intended to deal primarily with principles, a few standard methods of roughly obtaining quantitative evaluations of living things useful to the land manager are discussed briefly below.

Vegetation

Plant ecologists have devised numerous schemes for recording the species, density, and composition of vegetation. Most of them are careful methods of taking samples which may then be used as a basis for estimating the general condition of plant cover over an area which the samples represent. A given method also serves as a standard for comparing the vegetation of different areas (Weaver and Clements 1929).

QUADRATS

A quadrat is a sample area (usually a square, hence the name) of designated size in which the number and kinds of plants are recorded. Quadrats are frequently named with respect to the kind of record that is kept of them.

The *list* quadrat is one in which the number of individuals of a species is recorded. It is a census method, and is particularly useful in experimental work to determine percentage of germination, effects of disease, or similar results. The *clip* quadrat is much used by pasture and range managers. In its use, herbaceous vegetation is cut from a stated area at a particular height, as to simulate a certain degree of grazing. Oven-dry weight of the clipped plants serves as a standard of forage production and comparison. An actual plotting of plants occurring within a prescribed area produces the *chart* quadrat, a graphic record especially useful in yearly or other periodic analyses required to show detailed changes in vegetation. For meticulous work, large pantographs are sometimes used in preparing such charts.

In a *basal area* quadrat, employed primarily in forestry, the area occupied by the trunks of the trees, based on diameter at breast height (4.5 feet), is used as a measure of occurrence. This is usually expressed in square feet per acre. As an example, the occurrence of the dominant live trees in a Connecticut oak-chestnut forest in 1910, before the chestnut blight was introduced, was 217.4 trees per acre, with 84.23 square feet per acre basal area. After the blight, in 1924, there were 132.8 trees per acre, with a basal area of only 38.94 square feet per acre (Korstian and Stickel 1927).

Changes in vegetation can be most usefully recorded by regularly plotting the plants in *permanent* quadrats at stated intervals of time (Plate 8). Such quadrats, particularly if they are to express change due to variation in use, should be large. The effects of rodents, cattle grazing, or erosion can be much more correctly indicated by carefully observing and recording areas in acres rather than in square meters or feet. As an instrument of value to the land manager, the quadrat method of studying vegetation can be criticized because the samples are usually too small to provide a true picture of conditions. In so far as the quadrat is atypical, it follows that the more careful and minute the calculations based upon it, the greater will be the error. (Plate 8, p. 67.)

In order to learn the effects of burning, flooding, discing, or other management measures which subject an area to considerable disturbance, a land sample selected to record the results is sometimes spoken of as a *denuded* quadrat. On such areas many plants are destroyed, and studies of denuded quadrats are useful in learning something of the rate and nature of invasion and repopulation. They likewise serve to record the revegetation of protected range once depleted by overgrazing.

Photographs are sometimes taken as a record of vegetation, and thus we obtain a *photographic* quadrat, also spoken of as a camera set or tristat. Although such photographs are usually made from a camera tripod at a height and angle fixed and consistent for a given study, a novel and productive variation of the method was that used by Donald B. Lawrence in recording forest migration on newly formed volcanic slopes on Mt. St. Helens, Washington. Lawrence hung from a small balloon the camera with which, at a controlled height of 300 feet, he photographed areas several hundred feet square every five years. The resulting photographs have been referred to as balloon quadrats.

Attention should be given to the size of quadrat that can be expected to represent correctly the type of vegetation being sampled. How can one determine the smallest size of the individual quadrat and the smallest number of quadrats that will give a true picture of the plant cover? One way is to plot on graph paper the number of species (on the vertical axis) in a series of selected quadrats against the areas (on the horizontal axis) of the same quadrats and draw the resulting curve, called the species-area curve (Fig. 3). The point at which the curve flattens out parallel to the horizontal axis determines the size of area sufficient to represent the flora being studied. This is a graphic means of saying that when a sample area includes the species of most frequent occurrence, and all, or almost all, of the others, it is adequate to represent the vegetation type. The least number of quadrats can be determined in a similar way, by plotting number of species against number, instead of area, of quadrats.

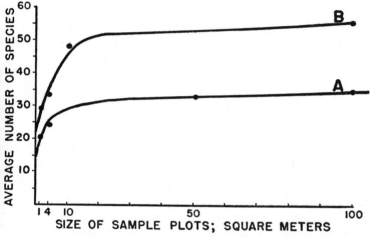

FIG. 3. Species-area curves. In plant community A the horizontal course of the curve starts at about 15 square meters, with about 30 species, which indicates a plot of that size will adequately represent the vegetation. In community B, where the average number of species is 52, a plot of about 20 square meters is necessary for an adequate sample (after Braun-Blanquet 1932, p. 55).

In a study of the red-maple swamp forest of central Long Island (Cain and Penfound 1938), it was found that 10 quadrats of 50 square meters each were sufficient to indicate the trees of the forest. Quadrats of 4 square meters were adequate for the shrubs; quadrats of 1 and 4 square meters for the herbaceous plants, and of 1 square decimeter for the mosses and non-flowering forms. Ten quadrats for each of these types of vegetation were usually sufficient to represent accurately the vegetation; otherwise 20 were used. Some method for determining size and number of samples, as well as the kind of sample itself, whether quadrat or other, will have to be considered by those who want to evaluate plant cover by the use of sample areas.

Another method of recording vegetation is the making of a transect. A transect is a cross section of vegetation recorded, tabulated, or charted in one way or another. Transects are named and used much like quadrats to record existing stand, plant succession, changes of vegetation due to use or other influence, and generally to serve as a concrete measure of plant cover. They are best adapted to show spatial changes in plant types along an expanse of terrain and are usually made at right angles to the extension of plant zones or ecotones.

A *line* transect is one in which there is recorded a list of plants or some diagram representing them, as they touch a string stretched along the ground. It is less descriptive than a *belt* transect, which represents a band of vegetation of determined width. A *belt* transect is essentially an elongated quadrat. It is much used in forestry, where a band of vegetation the width of a surveyor's chain (66 feet) is a customary standard. An interesting variation is the *photographic* transect to show density of plant cover, wherein a scale, placed on the ground, is photographed from a standard height and angle. A calculation based on that portion of the scale covered by vegetation serves as an index of vegetation density (Osborn 1940).

The *bisect*, in reality a vertical belt transect, is a cross section of vegetation as it is revealed by a trench deeper than the deepest root of the plants in the section. It shows, in a vertical plane, the roots of plants as well as the parts above ground. Bisects have been very useful in disclosing the composition of the underground parts of plants and the relation of the root system of one species to that of another. They have helped to prove that root competition is as influential in the occurrence of plant species as the relationship of the parts above ground.

CLOSURES

Whenever it is desired to learn something of the influence of animals upon plants, as grazing cattle upon range grasses, a fence is frequently built around an area to determine what the vegetation might be like if the animals were not present. On the other hand, the area may confine cattle or other animals in order to determine directly their effect upon vegetation. Such areas are frequently named with respect to the animals being studied; if the animals are excluded from an area, it is called an *exclosure*, but if the animals are confined within the fence, it is an *enclosure*.

Range managers have used closures throughout the western United States in an effort to determine the relation between grazing livestock and carrying capacity of the land, to check results of varying degrees of livestock grazing, and to seek an answer to the perplexing question of how much forage is eaten by jack rabbits and by rodents such as prairie dogs, ground squirrels, and kangaroo rats. By leaving a closure accessible to jack rabbits, excluding cattle by a fence, and removing grazing rodents by poisoning, it would seem an easy task to determine the effect of jack rabbits on vegetation. Although much has been learned this way, many of the studies are inconclusive, probably because the plots have usually been too small—rarely more than an acre in size, and usually not more than a rod square—to represent the complex of environmental conditions typical of large areas of range land. Furthermore, they are *controlled* experiments, with all the failings of such in biological research.

OTHER METHODS

The age of newly formed islands and floodplains, as well as the movement of trees into grassland, can sometimes be determined by *ring counts*, which indicate by the number of annual rings the age of the trees. Ring counts also suggest past rainfall cycles. By ingeniously correlating the pattern of annual rings in timbers from cliff dwellings of various ages, astronomers and botanists, working together, have been able to establish a time chart by which it is possible to date ancient civilizations of the arid Southwest.

The *relict* method of evaluating original conditions consists of seeking out undisturbed patches of vegetation in old graveyards, along fenced railroad rights of way, highways, or other places where native plant communities have been protected. Such protected spots preserve plant cover and associated animals somewhat as they existed before changes modified the community or caused it to disappear from unprotected places. They likewise preserve the original soil profile, giving us a standard against which to compute the degree of accelerated soil erosion that has occurred on disturbed areas. Over much of the United States, these remnants as they persist in sheltered spots are all we have to intimate the composition of original communities. It is unfortunate that we have not preserved intact even small areas of original vegetation representing at least major plant and animal communities. They would serve as invaluable standards against which to evaluate the land-management recommendations which the ecologist finds it his obligation to suggest.

Past conditions are indicated by *burn scars* on trees, which reveal times of heavy fires. They are likewise shown by blow-down injuries, stump sprouts, and other clues which the practical biologist should learn to recognize. *Pollen analysis* of bog deposits has revealed a great deal about forest cover in North America in postglacial time.

Mapping, of course, is as helpful in portraying specific environmental conditions, for example vegetative cover, as it is in graphically representing other landscape features. Aerial photographs, in normal times available for most of our country, are particularly valuable for mapping vegetation, especially when supplemented with sufficient ground surveys to determine correctly the composition of the types shown. The cover type maps commonly used to depict existing forest communities represent the acme of mapping technique as applied to vegetation.

The use of *experimental vegetation* should not be overlooked as a method of evaluation. A great deal has been learned about the physiological requirements of plants by carefully planned experiments. A knowledge of growth, reaction to light and other stimuli, nutrient deficiencies, and many other things of importance to those who use plants has come from experimental work. More recently experiments of transplantation from high altitude to sea level have taught us something new of plant taxonomy and ecology. The study of experimental watersheds promises to give us information about the effect of land use upon types of plant and animal communities.

Animals

It is more difficult to estimate populations of animals than to determine density and composition of plant cover for the simple reason that plants cannot hide. It is virtually impossible to find every individual of a given animal species on a prescribed tract of land, although ludicrous attempts are sometimes made to state deer, elk, and other big game numbers to the last unit in a figure of five digits for a state population. The best that can be said for animal census methods is that, provided a constant kind of sampling is used, it will give a comparative figure. This is no great pity, however, for management seldom depends upon an exact tabulation of an animal population. Brief descriptions of the more common methods that have been used for estimating animal numbers are given below.

The student interested in learning something about census methods will want to look at general summaries by Bell (1937), Studholme (1943), and Wight (1938). Lee R. Dice (1931a, 1938a, and 1941) and W. P. Taylor (1930) have summarized the ways of estimating numbers of mammals, while Dice (1930) and Lack (1937) have discussed census methods for birds.

MAMMALS

Actual counts of fair accuracy can be made to determine the numbers of some mammals, such as fur seals on their breeding grounds. Big game animals have been censused from airplanes traveling at a speed of less than 90 miles per hour and an altitude of about 500 feet. It is believed that antelope can be rather roughly counted in this manner, as can elk in winter when trees are devoid of leaves and the animals are conspicuous against the snow. Soper (1941) considered the airplane method a failure, however, in attempts to determine numbers of bison in the Wood Buffalo Park.

Attempts to count deer and other large mammals have been made by drives, in which a crew of men moves through the woods between two other lines of men forming the open end of a large V. As the crew moves forward into the closed end of the V, the animals running ahead of the crew are finally concentrated in a small area where they can be counted. The drives are usually made in representative forest or cover types over a stated number of acres or square miles, and total populations are accordingly estimated for the whole of the types represented. That an animal census at best is an approximation is shown by the fact that in an 800-acre deer pasture in rather open, rolling sagebrush-juniper cover in Utah, a careful census of the number of deer by drive counts listed only 45 per cent of the actual number present, as revealed later by a fire which killed all the deer in the enclosure.

One of the methods used to determine the number of beavers in an area consists of counting the number of houses plus bank burrows before the litters of the year have left home, and multiplying by five, the average number in a family. Careful observation at dusk, when the beavers are active, is considered to give a dependable count. The number trapped alive (so that they can be safely released) on a sample stretch of stream, related to the number of houses and burrows on the same stretch, provides a mathematical factor that has been used to determine beaver numbers over larger lengths of stream.

For jack rabbits, calculation based on the number within a strip of stated width along a road as observed from a car driven at a determined rate of speed—10 to 20 miles per hour—gives an estimate of numbers. Pellet counts have been used as an index of jack rabbit numbers, as they have for estimating cottontail populations. Counts of deer pellets have been made in order to learn something of deer wanderings and the type of forest utilized as well as the number of animals.

Small mammal populations are sometimes suggested by the number and species caught on a trap line of certain length over a given period of time, which provides a kind of trap transect. For territory typical of the transect, numbers may be roughly estimated for large areas. Trap quadrats are also used for the same purpose. Mammals can be tracked easily in the snow and in this manner a good deal can be learned of their ranges and numbers.

We know that, except when they have young, kangaroo rats average one per mound of earth characteristically thrown up by these rodents; their numbers can therefore be determined by counting the mounds of earth they upheave. Oddly enough, there has been so little interest in the study of animal populations in this country that such a simple method of estimating numbers of burrowing animals is not generally useful, for we do not know the exact relation between number of burrows and number of animals for even such common rodents as prairie dogs, ground squirrels, and pocket gophers.

BIRDS

For small birds, a count of singing males during the nesting season is indicative of total numbers, with the qualification, based on observation, that 40 to 60 per cent of the males are unmated. A count of occupied nests is a more reliable indicator of abundance. Colonial nesting birds can be enumerated by counting the nests and multiplying by two. Individual counts, recording only individuals seen or heard, unlike the song and nest censuses, are useful in both summer and winter. Ring-necked pheasants have been enumerated by the kind of 'windshield reconnaissance' mentioned above for jack rabbits. Pheasants and quail have been censused by drives in which dogs are used to flush the birds ahead of a moving line of men. This method is frequently used for game birds, with or without the use of dogs, or by a man and a bird dog. Birds with limited covey range, such as quail, are also counted by determining the number of coveys in an area and multiplying by an average number per covey. Sample areas are generally used as a basis for estimating total populations; time and weather are usually stated in reporting the counts.

Migratory waterfowl are sometimes estimated by dividing a flying flock into imaginary units, counting the number in one unit as accurately as possible, and multiplying this figure by the assumed number of units in the flock (Plate 8). When ducks or geese are on a water area, the area occupied is estimated in square yards and, at one bird per square yard, or whatever the average density appears to be, the total number of birds can be estimated.

Proportional methods of various sorts have been used to determine numbers of migratory waterfowl. One of the best is that devised by Frederick C. Lincoln (1930), which is based upon the number of ducks banded and the number of these killed during the succeeding hunting season. The total number of banded ducks killed during their first season as band carriers is to the total num-

---

ber of ducks banded as the number killed and reported by sportsmen is to the total duck population of that year. For example, if 600 banded ducks were killed during the first season after 5000 had been banded, and 5,000,000 ducks were killed and reported by sportsmen, the proportion would be 600 : 5000 :: 5,000,000 : x, where x is the total population for the year, in this case about 42,000,000. Although this method relies upon banding, reports of banded birds killed, and records of total birds killed by sportsmen, it is one of the most accurate means of estimating relative bird numbers that has been developed.

A method of estimating numbers of ruffed grouse much in use is based upon a scheme developed by Ralph T. King. The area to be censused is laid off in 40-acre blocks by compass lines. All the 40 lines and half the boundaries are followed by a man and dog, and the flushed birds recorded. The area of the transect covered (twice the average flushing distance times length) is to the total area in a particular cover type, density, and age class, as the number of birds flushed is to the total population of the same unit (Ashman et al. 1936). This may be expressed as an equation:

$$\frac{\text{Total area} \times \text{birds flushed}}{\text{Area of strip covered}} = \text{Total population}$$

PLATE 10

TOP. Class IV land in South Carolina. On this moderately eroded 13 per cent slope an intertilled crop may be grown one year out of many when, during the long rotation, grasses and legumes form the cover. Best use will consist of converting the land eventually into permanent pasture.

CENTER. Class V land (foreground) in the mountains of Colorado. Too poorly drained to be suitable for cultivation, it can provide excellent hay or be intensively grazed. The shallow, gravelly slopes in the background represent Class VII land, which requires severe restriction in use.

BOTTOM. Class VI land (foreground) in Colorado, which requires some restriction in use, as rotation grazing, and the application of conservation practices, as contour furrows, to maintain good grass cover and protect the soil. The rough land in the background is Class VII. (See Plate 11, p. 98.)

Shore birds have been censused by photographing a measured portion of a colony and counting the birds on the print. Reports of game bags or catches also indicate populations, especially if available for a series of years, and are useful, like furbearer trapping records, in determining cyclic fluctuation when the records extend over a sufficient period of time.

FISH

For determining numbers of fish, a so-called creel census is frequently used. This is simply a tabulation of fish caught as reported by fishermen for a given time and place. Number of fishermen, weight and kinds of fish, number of hours fished, average numbers of fish caught per hour, fish caught per fisherman, and number of hours fished per fisherman are sometimes enumerated. Failures, that is, the percentage of anglers who caught no fish, may also be recorded. Frequently, however, number of anglers, fish caught, hours fished, and catch per hour are the only data tabulated.

Fish in streams have been estimated by isolating representative sections of streams with barrier seines and counting the isolated fish removed by seining. To determine the number of fish in a pond, the fish are sometimes poisoned with a chemical, such as copper sulfate, or a plant poison, such as rotenone. Of course, the fish are thus killed, and such a census method is most useful for experimental work. In determining runs of migratory fish, traps of various kinds have been devised for holding fish until they can be counted, then released. A scheme comparable to Lincoln's method of censusing waterfowl is used for fish, wherein a stated number of fish are seined, tagged, and released. The proportion of tagged fish caught later to the total number of all fish caught provides a factor for estimating the total population. As with birds, information on numbers as well as movement, migration, and habits has been accumulated by catching, tagging, and releasing fish, which are later reported to authorities when recaught.

THE STUDY OF LIVING THINGS

It is very important that farmers and others be informed in advance of the likelihood of insect-pest outbreaks, as of upsurges in populations of other animals, and the United States Bureau of Entomology and Plant Quarantine publishes annually for the succeeding year a prediction of populations of grasshoppers and the Mormon cricket. These statements of expected abundance are sometimes accompanied by maps showing approximate areas of heavy, moderate, and light numbers. Analyses of the current numbers of the chinch bug, European corn borer, Hessian fly, and some other insects harmful to crops are also prepared and assist in evaluating possible future outbreaks.

Methods of estimating expected abundance vary, but frequently are based upon sample counts of egg numbers, as with grasshoppers, or upon estimates of numbers of hibernating adults, as with the chinch bug. In some cases, host plants are examined for general abundance of active individuals as for certain scale insects, or larvae may be counted to determine numbers, as they are for the codling moth. Traps of various sorts have been devised for this purpose. Although not strictly as a census technique, the airplane, as early as 1919, was used to locate cotton fields attacked by pink boll worm and to find mosquito breeding areas. Recently planes have been used to locate concentrations of Mormon crickets.

Sweeps with an insect net are used in estimating numbers of some insects, as the alfalfa weevil. A standard is 100 sweeps with a 10-inch net. The insect is not considered absent unless 1000 sweeps fail to obtain any. Cotton squares or bolls are frequently counted to determine the number of cotton boll weevils, with 100 or 200 squares examined at a stated number of places per acre. Insects inhabiting soil or forest duff are separated and enumerated by sifting samples through a series of graduated sieves. The abundance of some insects, as thrips, is indicated by the proportion of leaf areas discolored by them.

[85]

Whatever the samples may be, a definite unit for tabulation is used, as square foot, square rod, 100 plants, 100 fruits, or inch of bark or leaf surface, and the samples are taken at definite space intervals. Time, weather, and other conditions under which the census is taken are stated. Thus entomologists have developed many ingenious methods of calculating approximate abundance of specific insects, one of the most difficult groups of animals to evaluate in terms of numbers (Hyslop et al. 1925).

LANDSCAPES AND THEIR MEANING

THERE are many ways to study the land, depending upon one's purpose. The geographer is concerned with what he terms landscape, which to him usually connotes an entire panorama as viewed from the top of a high knoll—sky, horizon, forested hills, cultivated valleys, lakes, towns, moving trains, and all. The physiographer ignorés man's influence and concerns himself with the natural topography and configuration of an area. The agriculturist sees the cultivated fields and pastures; the economist the mills, roads, and towns; the sociologist the people and the culture they reflect.

For our purpose, a more intimate view of land is necessary. Not the detailed inspection of the micro-biologist or soil analyst, but a consideration of landscape as an ecological complex prescribing the use which man can make of the land. This involves, before all else, land classification. Just as the preceding chapter dealt with an analysis of plant communities and animal populations on the land, the present one treats briefly the classification of the land itself as an environmental influence, with land use especially in mind. Such treatment involves consideration of climatic factors as they relate to crop production, grazing, forestry, and other land uses, and it includes an examination of soil characteristics and types, especially with reference to productivity, slope, and degree of erosion. Other factors to be weighed in rating the use of land are water supply, size of areas uniform in character, location in the economic and social sense (with respect to towns, markets, etc.), and original vegetation. Land-use classification should relate to the physical capacity of the land to produce given crops for an indefinite period without exhaustion or waste of the land resource. It does not hold that production of tilled crops is the highest use of the land, but rather that this is true only for certain classes of land. Some land, for example swampy areas, is poorly used if it is not devoted to muskrat or other wildlife production, for efforts to produce more intensive crops only result in a waste of time, labor, and materials.

As Sauer (1921) pointed out more than 20 years ago, 'we desire to know for our lands the expectations of most permanent return under most complete utilization.' Sauer also contended that land classification was 'a qualitative grouping of the land, considering the land as a permanent resource under a normally developing economic system.' This view necessitates no more attention to the present use of land than to its possibilities of use. Likewise, it does not consider money value as a reliable criterion of classification, for where there is much wild land, even though it be fertile, the price of land is low, while near cities, regardless of its inherent productivity, land commands a very high price.

Anyone who has traveled across the broad expanse of our country, and who looks upon it as home, must sometimes feel despair at what he sees. Roadsides cluttered with cheap signs, unkempt streets, claptrap houses, littered backyards, streams greasy with industrial waste and sewage offal, once-rich farmland incised by eroding gullies, weedy range, grazed woodlots, and burned forests! It is not a pretty picture; we do not talk about it much; we are not proud of it. It is *not* America, the beautiful. Perhaps in the bounteous past we thought we could afford to ignore these things; yet we cannot. We need a kinship with the land—a way of knowing it, caring for it, and taking with gentle hands its friendly harvest.

The English agronomist, R. G. Stapledon, in the opening paragraph of his book, *The Land: Now and Tomorrow*, feels this need even for the long-settled British Isles when he writes (1935):

THE STUDY OF THE LAND

The culture of a nation by general consent, would, I suppose, be regarded as its greatest heritage, but a heritage perhaps equally worthy of being cherished is the land surface which a nation occupies. The culture to a large extent must have been influenced by the character of the land surface, and in any event culture and land surface are interwoven, and interact in countless directions difficult to unravel.

Without thoughtful use and management of the land, a nation cannot long endure.

TYPES OF LAND

As background for consideration of land, it is helpful to know something of how the earth's surface appeared before man transformed so much of it. Inasmuch as vegetation gives the most expressive ecological stamp to landscapes, the following tabulation is presented (James 1935):

LAND AREA IN THE WORLD

	PER CENT
Tropical forest	13
Boreal forest	9
Mid-latitude mixed forest	7
Mediterranean scrub forest	1
Dry lands	17
Grasslands	19
Mountain	18
Polar	16
	100

According to these figures, about 30 per cent of the world's land surface was covered with vegetation of some kind of tree form, 19 per cent was in grassland, 17 per cent in desert, and the remaining 34 per cent in mountain and polar areas.

Along this line, C. B. Fawcett (1930) made an interesting calculation to determine the amount of cultivable land in the world. Recognizing that not more than 12 per cent of the earth's surface has been surveyed from this point of view, he roughly estimated that of the 56 million square miles of land surface, 22 million, or about 40 per cent, are deserts, either too cold or too dry to support crops. Of the remaining 60 per cent, one-half consists of poor grazing land, forest, marsh, waste, or high mountains. The remaining 30 per cent of the land surface may be considered cultivable, that is capable of producing the ordinary crops of agriculture. At first it may seem that these figures differ radically from those in the preceding paragraph. However, if it is noted that the deserts of Fawcett's figures embrace polar areas as well as some dry and mountain lands as James tabulates them, and when other like compensations are granted, the two entirely separate calculations may be reasonably reconciled. Incidentally, Fawcett considered that not more than 2 to 5 per cent of all the desert areas could be made cultivable by irrigation. To the land manager, the significant thing is that so much of the earth's surface is desert—incapable of supporting cultivated crops, trees, or livestock—and that less than a third of the land is, in the ordinary sense, adapted to crop production.

Shantz (1941b) has calculated that 16 billion acres of the world exist in a crop climate (an area larger than Fawcett's cultivable land figure by more than 50 per cent!), which provides only 8 acres of potential crop land per present inhabitant. Considering the statements made by students of erosion (Jacks and Whyte 1939) that large portions of the United States, Australia, and South Africa are already impoverished by soil washing, one can easily wonder what portion of his 8 acres remains to support him. Such generalities, of course, are frequently all too meaningless, although they are thought provoking; of more immediate concern is what is being done to make the best use of the land we possess.

Of particular interest to our purpose are figures for the United States. The natural vegetation has been delimited on numerous maps and in the *Atlas of American Agriculture* it is stated that 48 per cent of the United States was originally forested, 38 per cent was in grassland, and 14 per cent in desert. Although this division is a very rough one, lumping all vegetation types into three great groups, it nevertheless emphasizes the fact that much of our country was once forested, that we had more than our share of the world's grassland, and that we originally possessed far less than our fair proportion of its desert area.

In the United States proper, there is a land surface of 2,973,700 square miles. More than 53,000 square miles of water surface—lakes, ponds, reservoirs—bring the total area of the country to more than 3,000,000 square miles. Expressed in acreage, it is 1,903,168,000 acres, and according to a recent compilation (Van Dersal 1943) of major types of land use, the acreage in the United States is apportioned approximately as indicated in Table 2.

TABLE 2. MAJOR TYPES OF LAND USE IN THE UNITED STATES

USE	ACRES	PER CENT
Farms	1,054,000,000	56
Forests		
National	155,000,000 ⎫	
Private (excluding farm woodlots)	216,000,000 ⎬	20
State	7,000,000 ⎭	
Public Grazing Land	182,000,000	10
Indian Land	55,000,000	3
Parks		
National	13,000,000 ⎫	
State	4,000,000 ⎭	1
Wildlife Refuges		
National	14,000,000 ⎫	
State	50,000,000 ⎭	3
Highways and Roads	20,000,000 ⎫	
Railroad Rights of Way	4,000,000 ⎪	
Cities and Towns	10,000,000 ⎬	7
Other	119,168,000 ⎭	
Total	1,903,168,000	100

These figures show that more than half the land surface of the country is in farms. A farm is defined as land which is directly husbanded by one person and his assistants, whether ranch, hatchery, farmstead, or greenhouse. Therefore, a farm includes many woodlots and much of the western range under private ownership.

[91]

If one considers land devoted to agriculture in its general sense, that is all farm and ranch land, including public domain, waste areas, and much of our wooded lands which are regularly grazed, it can be said that about 85 per cent of the land surface of the United States is given to agricultural pursuits.

More than one-fourth, about 27 per cent, of America's land surface is now government-owned—Federal, State, county, and city— the remaining 73 per cent is privately owned. It is surprising to realize that in the United States some 20,000,000 acres are already occupied by major roads and highways. The acreages that we must in the future devote to air fields, expanding urban areas, parks, and recreation centers will be no mean proportion of our total land surface.

Land-Use Classifications

Even in the early days of our history as a nation, attempts to classify land were made, though they consisted only of differentiating between land fit for agricultural, that is tilled-crop, use, and non-agricultural land. When the system of National Forests was set up in 1906, mapping and classification resulted in a separation of the surveyed lands into areas fit primarily for grazing and those suitable for timber. Gradually there developed a distinction between 'development' and 'conservation,' which was essentially the difference between immediate and permanent use of the land.

Early reports of the Reclamation Service distinguished areas of 'reclaimable' desert and cut-over forest or wild stump land, of which there was so much at that time. The reports indicated that the cost of making swamps and deserts productive of agricultural crops was even greater than that of clearing stump land. They also disclosed that the bulk of the land best suited to agriculture was in the humid eastern part of the country, and that only a small part of the stump lands were fit for agricultural development, a statement verified by subsequent experience. In 1921, H. H. Bennett presented a classification of forest and farm lands for the southern states. By 1924, the *Yearbook of the United States Department of Agriculture* included the report of a special committee, which concluded that 'the areas that are to be devoted to reforestation, as well as the areas that should be reserved . . . for pasture and crops, should be determined by deliberate selection . . . To this end, a systematic classification of our reserve land area is requisite' (Gray et al. 1924).

These early land classifications dealt with major uses of land on a broad scale, and it was not until rather recently—within the past ten years—that classifications of land applicable to the planning and management of a single operating unit, such as an individual farm or ranch, have been undertaken.

It should be emphasized that the land inventory, which usually consists of mapping soils, vegetation, or other features of the landscape, differs from land classification and planning. Early work consisted primarily of making an inventory of land conditions. Land classification, however, implies the making of specific plans for the land and usually includes recommendations for its use. As P. S. Lovejoy remarked in an early essay on the history of the classification idea in America (1925), 'Classification and land planning go hand in hand, for classification is essentially purposive; it looks to the attainment of some end and hence includes planning.' Land classification as considered in this book, and in its modern sense, combines inventory, classification, and planning. A final step is to transform the plans into operations on the land. There are many ways of accomplishing this transformation, and, although they are phases of human endeavor largely beyond the scope of the present work, they will be touched upon.

In Chicago in 1931, a National Conference on Land Utilization was held. Interest in land classification took concrete expression at the first National Conference on Land Classification, which convened at Columbia, Missouri, in 1940. At the latter meeting, land classification was discussed with respect to soil and its development, and as it relates to soil classification; it was considered in connection with rural zoning, as an appraisal and credit aid, and as a guide for real-estate development. At the conference, the classification of land was further weighed as an implement for

[93]

facilitating the management of forests and grazing lands, the establishment of soil conservation practices, and general farm management; also for planning recreation areas and reclamation projects, and for general land-use planning. A summary of various schemes of classifying land in the United States has recently been issued by the National Resources Planning Board (1941). It is apparent that much is now expected from land classification, and such an interest is recognition of the fact that the land, more than any other natural resource, is the foundation upon which we must build our future development as a nation, if we are to manage our affairs in a sound, reasonable fashion.

Just as there are many aspects to land classification, so there are many ways in which a classification of land can be developed. One of the systems of evaluating land is the productivity index or productivity rating (Ableiter 1937), based primarily on the yield and quality of crops under physically defined systems of management. Another system is the Storie index (Storie 1933). Based upon soil profile, soil texture, and other physical modifying factors, this index is a 'numerical expression of the degree to which a particular soil presents conditions favorable for plant growth and crop production under good environmental conditions.' Still others are based primarily upon the use being made of the land. An example is that developed by the United States Geological Survey for the Great Plains, which classified land into (1) tillable land, subdivided into irrigated, irrigable, dry farming, farming-grazing, and grazing-forage land, and (2) non-tillable grazing land.

Numerous states have classified land for purposes of management, using various criteria. A few examples will illustrate.

In 1920, the State of Michigan began a survey of idle land on a basis previously little used. Actually three simultaneous surveys were combined to afford a land classification. The surveys evaluated (1) civil base data—topography, cover (vegetation), and present use; (2) soils; and (3) census, tax, and related economic data (Schoenmann 1923; Veatch 1933). In Wisconsin, land classification was also started early, and there a further important step was taken, for Oneida County, in 1923, was the first county in the United States to regulate rural land with respect to agricultural, foresty, and recreational uses. In Wisconsin, the criteria used for classifying and planning use of the land were: (1) tax delinquency, (2) location of farms, both operating and abandoned, (3) land ownership and land entered under the State forest crop law, (4) location of schools, school district boundaries, and school bus lines, and (5) soil maps showing main soil types (Whitson 1935).

In 1933, the Minnesota State Legislature created a permanent State Land Use Committee, and a land classification subsequently developed by that state divided rural land into two classes, namely, conservation zones devoted to timber and other conservation purposes, and agricultural zones (Jesness et al. 1935). This system was developed in considerable detail for 14 northeastern Minnesota counties. Determination of zones—conservation or agricultural—was based on assessed improvements and settlement pattern as they related to soils, roads, schools, and cover. Maps based on these factors were then revised by various county officials, and the zone boundaries were checked in the field both by inspection and by interviews with residents. In 1934, the Land Use Committee of the Washington State Planning Council issued a report along these lines and other states have more recently given attention to the subject.

Michigan, Wisconsin, and Minnesota developed rural zones on the premise that land should be classified on a broad scale, and that zones of use should be delimited for large areas rather than specific parcels of land. As the need for more specific rural zones or land classes became apparent, schemes for classifying land more precisely were developed. Two state classifications of a detailed type are briefly stated below.

Prefacing the consideration of land upon the fact that present land use is the result of a long period of experimentation by farmers in attempting to find the most profitable uses for lands of different character, New York State classified land to determine areas which should be dedicated to forests and recreation and the farm lands to which it was advisable to extend rural electrification (LaMont 1937). This classification was based primarily on a survey that evaluated the size and condition of farm buildings and the apparent amount of business being done, although some consideration was given to present land use, soil, elevation, topography, and probable future use. The scheme depended upon both past and present use, so that land that once had been cleared for cultivation, but later had been abandoned and was reverting to trees when the survey was made, was considered to be best adapted to forests. Old stump fences were taken to suggest suitability of land for the production of trees, which the stumps indicated must once have occurred there.

The New York classification grouped lands into eight classes. Class I consisted of abandoned farms, idle lands, or forests, for which the recommended use was forest and recreation. Class II consisted of idle, abandoned, or poor operating farms, which it was believed could most profitably be devoted to forest and recreational purposes. Classes III to VII inclusive were lands considered to be best suited for permanent agriculture, the more intensively used the higher the number. An additional class included the residential areas, such as urban, suburban, or village tracts, and areas used for industrial or commercial purposes. Use of the classification is illustrated by the fact that rural electric service was recommended for Classes III to VII.

Attempts to classify rural land with respect to the most advisable use for particular parcels of land in order to conserve soil and maintain its fertility, and to recommend the production of crops, pasture, or forest on areas best adapted to them, is shown by the classification developed for the State of Tennessee (Hendricks 1936). Ten classes were recognized, the first four suitable for cultivation with decreasing intensity of use. Class 1, for example, could be most intensively used, because it had no fertility or erosion problem; Class 4 required 7-10 year crop rotations in order to maintain fertility and hold the soil. Classes 5 and 6 represented pasture land, the latter being crop or idle land that should be converted to pasture in order to maintain permanently its productive condition. Class 7 was land that must be converted to forests. Class 8 was already in forests and should so remain, while Class 9 was low land that, if drained, could be converted into Class 1 or Class 2 land and be permanently cultivated. Class 10 was waste land and lands in roads, farmsteads, and so on.

The examples above show that some classifications of land strongly rely upon past or present use, even reflecting in some instances the skill of the individuals operating the land. Others are attempts to determine the most advisable management irrespective of the use that has been or is now being made of the land. It is undesirable, of course, to think of recommendations for land management without considering the needs and problems of the people who use the land. Land managers need a classification which portrays the physical capacity of land to produce over a long period of time under stated conditions of use, and which can provide land operators with a basis for actual practice on specific parcels or units of land. Such a scheme is the Classification of Land According to Use Capabilities.

THE CLASSIFICATION OF LAND ACCORDING TO USE CAPABILITIES

Although social and economic influences greatly affect the use of land, it is becoming generally recognized that the classification of land should be based on its natural characteristics rather than upon the skill of the individual operating it or upon prevailing economic conditions. This is essential if the classification is to serve as a basis for the most intensive sustained use consistent with preservation of the land as a permanent productive resource. The ultimate purpose, of course, is to sustain at a permanently high standard of living the people who now live on and those who in the future will occupy the land. Classes of land according to use capability are determined by considering the physical factors influencing land use. These factors include soil conditions—physical, chemical, and biological—slope, kind and degree of erosion, and

[97]

NATURAL PRINCIPLES OF LAND USE

TABLE 3. CLASSES OF LAND ACCORDING TO USE CAPABILITIES

A. *Suitable for Cultivation* *

  I Without special practices (fertilizers and simple crop rotations may
     be used)
 II With simple practices (such as contour cultivation, strip cropping, or
     simple terrace systems)
III With complex or intensive practices (as terraces; usually a combina-
     tion of practices is needed)
 IV With intensive practices and limited use (long rotations with no in-
     tertilled crops, or cultivated in small acreages)

B. *Not Suitable for Cultivation; Suitable for Permanent
Pasture (Range) or Woodland*

  V Without special practices; land only slightly susceptible to deteriora-
     tion (grazing of range to full carrying capacity; cutting of forests
     without special practices to protect the land)
 VI With moderate restriction in use; land moderately susceptible to
     deterioration (rotation grazing; logging with careful location of
     trails and other practices to protect soil)
VII With severe restriction in use; land highly susceptible to deteriora-
     tion (on range only occasional grazing; in forests only highly selec-
     tive logging)

C. *Not Suitable for Cultivation, Pasture, or Woodland;
Suitable for Wildlife*

VIII With or without special practices (productive of useful wild plants,
      furbearers, game birds, fish, and generally serves as wild animal
      range)

 * Cultivation means tillage of the soil as practiced with intertilled crops
and in preparing land for grain.

PLATE 11

TOP. Class VII land in Vermont, devoted to timber production. Forest management on such lands requires intensive practices, as highly selective cutting,
if yields are to be maintained over a long period of years and the soil remain
permanently productive.
BOTTOM. Class VIII land in Pennsylvania. Lands of this class are not permanently productive of cultivated crops, livestock, or timber, but frequently
yield highly valuable crops of wild plants or animals. This marsh provides an
economic return in the form of muskrat pelts.

[98]

certain other environmental features, such as climate and drainage.

Accordingly, eight classes of land are recognized (Norton 1940; Hockensmith and Steele 1943), as shown in Table 3. (Plates 9, 10, and 11). It should be noted that this classification refers to rural land, to the use of which ecological principles are particularly applicable. Urban, industrial, airport, and recreational areas, of course, should also be established whenever feasible with reference to a scheme of land classification. (Plates 9, 10, 11, pp. 82-3, 98.)

As an illustration of the factors considered in determining a class of land, let us briefly describe Class 1 land. It is, first of all, land highly suitable for cultivation, for it does not have a permanently high water table; neither is it stony or spotted with rock ledges; nor does it possess any other physical characteristics which interfere with the use of tillage implements. Furthermore, clean-tilled crops like corn, cotton, or tobacco, the growing of which is often likely to result in soil washing, can be raised on this land without danger of appreciable accelerated erosion. Finally, it retains and supplies sufficient moisture and plant nutrients to maintain those physical, chemical, and biological conditions of the soil that favor continued production of moderate to high yields of farm crops.

This example serves to indicate that the classification of land according to use capabilities is based upon the capacity or potentiality of land to produce. It does not necessarily reflect the present status of a land area, which may need the application of conservation practices, drainage, or other treatment to bring it

---

PLATE 12

TOP. Man-induced erosion is solemnly depicted in this community cemetery near Chester, Ohio. For more than 100 years the plot has been fenced. It now stands three feet above the surrounding land, worn down by use and mismanagement.

BOTTOM. To prevent accelerated soil removal, many special practices are being applied to the land. On cultivated fields in New Mexico, contour furrows retain snow moisture, prevent rapid runoff, and break the flow of silt-laden waters across the long slopes.

[99]

into the condition which will make it most permanently productive. On the other hand, if drainage and conservation practices are not able to transform an area into a more permanently productive condition, that land must remain in a less useful class.

There are, of course, many localities in which land may not be as completely utilized as indicated by a classification according to use capabilities. Such a classification determines the most intensive use to which parcels of land can be subjected without waste of the land resource. Social and economic conditions always influence use of land. Although much of New England is Class ii and Class iii land, capable of producing cultivated crops, it is now most profitably maintained in pasture and forest, because tilled crops from these lands cannot compete with similar products from even richer crop lands like those of the Middle West. The wooded hills of Massachusetts were at one time four-fifths in farm land, and the grass-covered valleys of Vermont once grew so much wheat that the State was known as the Bread Basket of the Revolution.

In the following chapters, the one dealing with Farms is concerned largely with land Classes i to iv; the chapters on Range and Forests deal with Classes v to vii; and that on Wildlife primarily, but not exclusively, with Class viii land. The chapter on Waters deals in part with Class viii land also, because in so far as their profitable management is concerned, streams, ponds, and lakes should be looked upon as Class viii 'land,' for they are best adapted to the production of a crop of wild plant and animal life.

Such an attempt to classify land must be more than a 'paper idea'; it is imperative that some such scheme be adopted if we are to use each parcel of land productively for the purpose to which it is best suited, and preserve it for the future. Enough crop land in the United States has already been totally lost to ordinary farming through erosion—an acreage equivalent to the combined areas of Texas and California—to prove that some design must be patterned to prevent misuse of soil. Furthermore, the classification outlined above has already served as a basis for land-use operations. On more than a quarter of a million individual farms and ranches throughout the United States, from the dairylands of New England to the cotton plantations of Alabama, from the cornfields of Iowa to the citrus groves of California, from the wheat lands of the Palouse to the ranges of west Texas, farmers co-operating with the United States Department of Agriculture have readjusted their use of the land to check soil blowing, reduce soil washing, and bring their land-management practices into agreement with physical and environmental capacities of the land, until in many sections the agricultural land-use pattern has been completely revamped.

In this new pattern, crops lie in strips across the slopes of the fields; furrows follow the contour of the land, as do orchards and hedges; eroding field borders and odd spots are clothed with a protective cover of useful perennials; pastures protect the gentle slopes; and woodlands cover the steep hills. What has been done, of course, is only a beginning. Only a fraction of the land has been carefully planned and many operations need improvement and revision. But already enough has been accomplished to suggest the shape of things to come in land management, and as the land pattern changes in response to man's thoughtful direction, its design will reflect the influence of those who have looked to the classification of land, and who have worked with land operators to make it a practical, productive scheme. Something more about the actual application of plans based upon the classification of land is written in the following chapter.

LAND-MANAGEMENT TOOLS

It can scarcely be overemphasized that any one who works on the land should relate his operations to the physical condition of the land, although in practice this has been far from axiomatic. Long years of working square fields on round hills developed a habit that retarded the acceptance of such a simple, fundamental idea as contour cultivation, a labor-saving as well as a soil- and water-conservation measure that is known to increase crop yields.

We must learn that, while some land can be used repeatedly for crop production, because of favorable type of soil, degree of slope, condition of erosion, and other physical characters, other specific areas are best suited and most profitably used for long rotations, permanent pasture, woodland, or production of wildlife and special crops such as cranberries or blueberries.

The land-management biologist must understand the reasons for determining the use to which each parcel of land is best fitted; he must familiarize himself with methods of management and objectives to be reached by the methods employed; and he must be prepared to determine means of integrating biological recommendations with those for the primary use of the land. Some of the means as well as some of the problems of integration are discussed in the remaining chapters.

If this book were not intended to treat ideas and concepts rather than methods and techniques, a great deal more space would be allotted to means of managing the land. An entire chapter would also be given to the tools of the land-management biologist which, while still imperfectly understood and little used, will someday rank with the techniques employed by land operators and others who have had long experience with the land. But we cannot neglect to mention some of the methods by which the practical biologist applies the particular principles with which he deals.

The *classification of land* discussed above is a primary tool of the land manager. Means of evaluating numbers of plants and animals listed in the preceding chapter are also invaluable tools. Throughout subsequent chapters it will be noted that examples of the application of natural principles frequently involve manipulation of vegetation, animals, and land use. These methods are essentially instruments of land management. Most of these tools are employed to modify some element of the habitat as man wishes it to be modified.

One of the agents most devastating to habitat is fire. It is not only destructive but, as we shall see, it frequently determines for decades, and even centuries to come, the type of vegetation and associated animals which follow severe burns. Fire has been employed to burn off undesirable plant cover for a long time, but its controlled use as a tool in land management is very recent. Fire is now an accepted tool in the management of bobwhite quail in the Gulf Coast States (Stoddard 1939), and it is used to burn off old, dry bunch grasses in Louisiana coastal marshes in order to make the young, tender leaves available as food for the snow and blue geese which winter there. Foresters are cautiously looking to fire as a management tool in the South, where stands of long-leaf and other important timber pines can be perpetuated by careful burning. In the West fire has been recommended as a management measure in ponderosa pine forests. When subjected to repeated burning, these forests reproduce themselves; when protected from fire for 30 to 40 years, they begin to be replaced by Douglas fir, incense cedar, and white fir (Weaver 1943). The controlled use of fire—that is, burning with low intensity at a time of day and year and under such other conditions as will result in predictable, desirable effects—is a technique about which we shall undoubtedly learn a great deal as the need for managing land becomes more acute.

One of the most important tools of the land manager is the *fence*. Conservationists have called fencing their most useful management practice. By means of the fence, grazing stock can be kept from woodlands, crops, ponds, streambanks, and other areas where they might disturb not only the desirable plants but, by reducing vegetation and by trampling, cause accelerated erosion. Land-use adjustments, such as converting side-hill corn fields to pasture, and pastured bottomlands to cultivation, are frequently made possible by fencing. The western range is becoming a network of large pastures, where rotation grazing is facilitated by the separation of the range into fenced areas.

Another land-management tool assuming more and more importance is *timing*. Not only by rotation grazing, but by deferred grazing, a pasture may be materially improved. Thus livestock are kept from a pasture until the grass has reached sufficient growth that grazing will not damage it. Even such a simple matter as the time of planting a crop may have considerable effect upon ultimate yields. We now know that late planting of corn, use of resistant hybrids, and clean plowing where erosion is not increased by such practice, combine to reduce greatly the damage by the European corn borer, one of the most destructive insect enemies.

Perhaps the most fundamental tool is land-use conversion. So long as a parcel of land is being misused, land-improvement measures are likely to be futile. Methods to control destructive wind erosion in the Dust Bowl are prohibitively costly and impermanent as long as so much of the land there unsuited to tilled crops is cultivated. When improperly cultivated areas are abandoned in favor of grazing, and an adaptable grass cover is established, the land is amenable to management and becomes a permanently productive resource. The major task of the land manager is to bring each acre of land into the use for which it is naturally best suited, so far as prevailing social and economic conditions permit. To do this will require many changes in present use, possible only after we have learned more than we now know about such matters and land operators have had a long time in which to accomplish the desirable adjustments.

Many other techniques are mentioned on subsequent pages. They include tools peculiar to particular technicians, as distribution of stock water tanks by the range manager to obtain uniform grazing, and the planting of trees by the forester. Control measures involve other tools, as poison baits for reducing injurious animals and chemical sprays for killing weeds. Whatever the tools, they can be used wisely or to no good purpose. The land-management biologist will find that the effectiveness of his work depends very much upon the success with which he uses land-management tools. To this end he must be familiar with such tools, of which there are many, learn what they can do, and relate their use to ends that are naturally sound. By such means, his judgments can contribute to wise and reasonable use of the land.

FIRST RESOURCE

IN a sense agriculture is the most exalted kind of land management, for in it we have consciously attempted to reduce the abundant plants of the land to a single species or variety which we wish to grow at a particular time and place. Not only is cultivation fostered on lands where competing plants and animals are eliminated as far as possible, but the tilled crops themselves represent plants so modified by man's attention that their botanical origins are often no longer known to us. The management of cultivated crops is accomplished only by great effort, constant attention to the land, and intensive use. Today that intensity of use is of growing concern, for we are learning that we must temper use of the land with sound management and balance it against the common good.

The wisdom of our forebears somehow eludes us, else many of our agricultural problems might have reached an earlier solution. This is brought home when we leaf through early American accounts of farming, such as the memoirs of the Philadelphia Society for Promoting Agriculture, a society founded close upon the molding of the nation 'by some citizens, only a few of whom were actually engaged in husbandry, but who were convinced of its necessity.' One of its members, Charles Caldwell, in 1808 wrote to the Society's president, the ever-inquiring Richard Peters, Esq., a long letter on numerous agricultural problems of the day, contending that 'agriculture forms the true basis of our national prosperity. Its enlightened and industrious patrons and promoters, therefore, are justly ranked among our soundest patriots.'

But the encouragement of these men was made of more than words. They realized that many of the farm practices of their day were injurious to the land. They wrote that the usual practice among farmers of the time 'consisted in a series of exhausting grain crops, with scarcely any interruption, for several years; after which, the land was abandoned to weeds and natural grass, under the fallacious idea of rest; and, when completely worn out, new land was cleared, and the same wretched system pursued.' With the conviction that the country could not long remain bountiful under such a system, the Society early set up premiums for promoting agriculture. In the year 1791, the Society offered, among many other material prizes for good husbandry, the following (the italics are mine):

For *the best experiment of a five years course of crops*—a piece of plate, of the value of two hundred dollars, inscribed with the name and the occasion; and for the experiment made of a like course of crops, next in merit—a piece of plate, likewise inscribed, of the value of one hundred dollars . . .

For the best information, the result of actual experience, for *preventing damage to crops by insects*; especially the Hessian-fly, the wheat-fly, or fly-weevil, the pea-bug, and the corn chinch-bug or fly—a gold medal; a silver medal for the second best . . .

For the greatest *quantity of ground, not less than one acre, well fenced, producing locust trees*, growing in 1791, from seed sown after April 5th, 1785; the trees to be of the sort used for posts and trunnels, and not fewer than 1500 per acre—a gold medal; for the second—a silver medal . . .

For the best method, within the power of common farmers, of *recovering old gullied fields* to a hearty state, and such uniformity, or evenness of surface, as will again render them fit for tillage; or where the gullies are so deep and numerous as to render such recovery impracticable, for the best method of improving them, by planting trees, or otherwise, so as to yield the improver a reasonable profit for his expenses therein, founded on experiment—a gold medal; and for the next best—a silver medal.

A century and a half has passed since then. The practices the Philadelphia Society encouraged seemed to become trivial in the face of an expanding, prosperous people. Yet anyone familiar with present efforts to assist land operators with agricultural problems must see in these early recommendations such a parallel with contemporary programs that it draws him up with a start. How long can we continue to neglect the needs of the land? Within the past few years, fortunately, there seems to have developed in the American people a real consciousness of the fact that the land needs care, and that the most basic of all resources, the soil, must be protected, managed, and wisely used.

Erosion

Although erosion occurs in all classes of land, and is as detrimental to overgrazed range and mismanaged forest as to cultivated acres, the subject is discussed in this chapter because accelerated erosion is conspicuous on agricultural land, where most spectacular methods have been applied for its correction. There is probably nothing associated with the use of most of our agricultural land, whether in crops, pasture, orchard, or woodlot, that is more destructive to the environment than accelerated or man-induced erosion (Bennett 1939; Jacks and Whyte 1939) (Plate 12). Furthermore, it is now so widespread throughout populated areas that it reflects itself in the social and economic welfare of nations.

With respect to their influence upon habitat, there is a significant difference between normal or geological erosion and man-made or accelerated soil removal. Erosion induced by man may cause unusually rapid and profound changes. Nevertheless, it has been much neglected by ecologists. Although grazing, lumbering, and fire resulting from the activities of man have long been recognized as environmental influences, it is only recently that textbooks on ecology have considered accelerated soil erosion as an important ecological process. This disregard is also well illustrated by the fact that until a few years ago many extensive agricultural field investigations were conducted with careful consideration of climatic factors, application of fertilizer, and crop yields, but with no consideration of erosion, which in many instances was removing soil from the land on which the experiments were being conducted at a rate sufficient to invalidate the results of the investigations. (Plate 12, p. 99.)

Modern theories of soil genesis acknowledge five influences in the production of natural soils: climate, native vegetation, underlying rocks (geology), relief or slope (physiography), and age. That soil is the result of the dependent interaction of these several factors replaces the older idea that it is simply decomposed rock—a result of destructive weathering—or results entirely from vegetation.

A given soil type, therefore, is the result of the continued interaction of several processes, including the physical disintegration and chemical decomposition of mineral constituents and the recombination of some of the resulting substances. Such substances, together with organic materials composed of plant and animal residues, are being constantly redistributed as a consequence of water movement, leaching, and the activities of the soil microflora, microfauna and, to a lesser extent, larger animals such as earthworms and burrowing rodents. These physical, chemical, and biological processes normally maintain a soil profile of definite depth and composition, usually of three well-defined, yet genetically related, horizontal layers, the A, B, and C horizons (Fig. 4). The A horizon is popularly called topsoil, the B subsoil. Together they constitute the solum, or true soil. Below them is the C horizon consisting of parent material that is weathered, but still unchanged by other soil-building processes. Underlying the C horizon is the unweathered and unchanged parent material.

In very broad terms, it may be said that normal erosion, a process as old as geological history, is in most regions of the world an imperceptibly slow, although persistent, action. It is usually so gradual that as upper portions of the soil profile are removed by wind or water, soil-building processes compensate for their removal by developing the lower portions of the profile. The result is the gradual downward progress of the entire profile and, geologically, eventual leveling of the landscape, or peneplanation.

One of the first effects of man-induced erosion is the removal of portions of the A horizon at a rate faster than formative processes can replace them—an effect often evidenced by sheet erosion and the formation of rills. If the process continues, sheet erosion may be followed by the removal of portions of the B horizon and

PROFILES OF THREE SOIL GROUPS
(DIAGRAMMATIC)

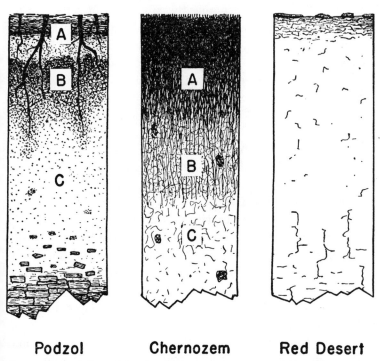

Podzol   Chernozem   Red Desert

FIG. 4. From left to right, three soil profiles typical of forest, grassland, and desert. The desert soil is without defined horizons, the others exhibit characteristic A, B, and C horizons.

the development of gullies. Gullies may erode into the C horizon and are known in many instances to have cut below it to remove portions of the parent material also. The loss of soil by such process is, on land under human occupation, serious and costly. To restore the loss may require a great many years of the most careful land management, for topsoil is not easily compensated for by the application of fertilizer, better seed, or improved equipment.

There are many differences between the A horizon of a soil and lower layers. Temperature, air content, and water-holding capacity vary with depth, and infiltration rates vary greatly for different horizons. Organic matter, as roughly expressed by the carbon-nitrogen ratio (a figure denoting the proportion of carbon to nitrogen in the soil), usually exhibits a steady decrease with depth of soil. The hydrogen-ion concentration and amounts of essential elements also vary with soil depth. As a general rule, the content of nitrogen in soils is greatest in the plow level. In the surface layer of many soils, approximately one-third of the phosphorus present may be in organic combinations, while in subsoil much less is in available form. The potash content of soils may not display a general correlation with soil depth, but because potassium remains essentially soluble and does not accumulate, as does nitrogen in proteins or phosphorus in bones, it can be easily lost by washing of the soil surface.

Life in the soil also varies with depth. The numbers of bacteria in humid regions decrease with depth of soil, and there is a corresponding correlation between certain soil bacteria and soil depth. Soil protozoans may be practically confined to the uppermost few inches of soil. Shrews, moles, burrowing rodents, and other vertebrates may work both the surface and subsurface layers but their activities are most effective in the upper horizons.

Jacot (1936) has emphasized the great importance of minute spiders and their kind in maintaining normal soil structure. His investigations show reduced populations of these small invertebrate animals with increase in soil depth. For example, 2,800 humus-eating mites per square foot were counted in the top 3 inches of soil, while 10 to 13 inches below the surface only 28 of these organisms occurred. Likewise, the number of such insects together with centipedes and millipedes dwindled from 500 per square foot 1 to 3 inches below the surface to 22 at a depth of 10 to 13 inches. Soil erosion eliminates the porous channeled layer in which these minute animals are most abundant, and repopulation of abandoned agricultural land by these organisms is slow.

Yields from crop plants express differences between the A horizon of a soil and lower layers. Experiments reveal that, under comparable conditions of slope, rainfall, and cultural treatment, the per-acre yield of many agricultural crops on subsoils is materially less than the yield produced on topsoils of the same soil types. Even with heavy applications of fertilizer, it is generally impossible to bring the yield from subsoil up to that from the related surface soil, so great is the difference in their composition. This difference is important, for in spite of higher quality seed, more productive varieties of crop plants, better fertilizers, and improved cultural practices, the average per-acre yield of cotton for the United States has declined materially, corn has not quite held its own, and even the yield of wheat has increased but slightly.

Of course, the reasons for such decline cannot be easily assigned. Increased use of farm machinery during the past several decades has emphasized greater production at the expense of acreage yield, and the extension of agriculture into regions less suitable to major crops has also tended to reduce production per acre. There is considerable evidence to indicate, however, that accelerated soil erosion is a factor that has contributed notably to the limitation of crop yields on agricultural lands.

Conservation Practices

Agricultural land is varied, and on many American farms there are pastures, woodland, and other uncultivated areas in addition to tilled fields. The ecological principles applicable to the management of non-cultivated lands are treated in subsequent chapters. On cultivated land, ecology becomes largely a matter of the relation between an individual plant—as a corn stalk—and specific environmental factors—as soil condition, cultural treatment, insect enemies, or length of day. There is comparatively little community influence among the artificially sustained plants of a field of tobacco. Thus the ecology of tilled acres is primarily physiological, and of less concern to the land-management biologist than that of most other types of land.

However, there is much that the biologist can contribute to the management of cropland. The development of a classification of rural land in accordance with its capacity to produce is in itself an ecological task of the first order, and the determination of the types of crops to be grown on various classes of cultivable land requires a broad knowledge of relationships. Furthermore, the determination of the kinds of conservation practices that will permit the most intensive use of each class of cropland consistent with its permanent productiveness is essentially an ecological problem. (Plate 13, p. 114.)

Even some of the old agricultural practices, like plowing, are now questioned. Ever since Jethro Tull in 1733 published his revolutionary book, *The Horse-hoing Husbandry: or, An Essay on the principles of tillage and vegetation*, not only turn plowing but repeated cultivation has been looked upon as a necessary step in the raising of tilled crops. The reader may have begun to wonder, as he read of erosion and inherent differences between topsoil and lower layers of the soil profile, whether plowing could be an unmitigated good when it so disturbs the soil.

In the Great Plains, disking has been employed not only as preparation of the soil for planting grain, but also to kill weeds which might use up soil moisture when the fields are fallow. Although disking does not turn over the soil as a moldboard plow does, like plowing it disturbs the soil and subjects it to erosion and leaching. As an experiment, grain stubble was left on the ground to prevent wind erosion. This checked the blowing somewhat, but the surface was still disturbed by the disking. It was then learned that the grain could be seeded in the stubble by drills without the usual disking of the soil. Furthermore, weeds can be destroyed, not by churning the top of the ground, but by cutting off their roots. This is accomplished by a special implement which draws a blade or rod beneath the surface, with comparatively little disturbance to the soil structure. (Plate 14, p. 115.)

Admittedly not perfected, and adapted to particular conditions, this development opens up the entire question of whether the severe disturbance caused by plowing in the conventional manner is prerequisite to production of agricultural crops, and whether new methods of cultivation should not be sought. If the surface of the soil were disturbed less, both wind and water erosion might not be the problem that they have been, and the soil, as a productive natural element, might be preserved with less severe change than it has suffered so often in the past.

Many conservation practices adapted to cropland are already well developed and aid immeasurably in preserving the soil as a valuable resource (Plate 12). Appropriate crop rotations, strip cropping, contour cultivation, terracing, field border plantings, vegetated terrace outlets as parts of water disposal systems for farm fields, and many others all tend to hold the soil and maintain its permanent productiveness. Let us look at a few of these practices that are of more than ordinary concern to the land-management biologist.

A soil conservation practice widely recommended in parts of the United States is strip cropping, which is the establishment of a clean-tilled crop, such as corn, cotton, or tobacco, in broad strips alternating with strips of a close-growing crop, such as alfalfa or meadow grasses. The strips are usually established on the contour, thus checking soil washing. The close-growing crops retard the flow of runoff water across the slope, reducing its erosive power. Many questions have been raised about the use of this practice, some of them biological.

Does the practice increase the number of injurious insects, for instance? Chinch bugs may winter over in a meadow strip to move the following spring into corn planted in an adjoining strip, and strip cropping may aggravate grasshopper damage to wheat when it is grown in strips alternating with strips of wheat stubble. Several students have dealt with various phases of this subject (Bishopp 1938; Strong 1938; Harris 1939; Annand 1940; and Wilbur, Fritz, and Painter 1942).

There is reason to believe, however, that strip cropping may increase the effective control exercised by predacious insects upon harmful species (Marcovitch 1935). Likewise, bird populations, increased by strip cropping, may to some extent assist in controlling insects. Studies in Ohio have shown that in fields composed of strips of corn, small grain, and meadow, there were approximately twice the number of breeding ground-nesting birds per acre as in comparable nonstripped fields (Dambach and Good 1940). The land-management biologist can assist materially in weighing the damage caused by insects against the usefulness of birds and the significance of the result in terms of its influence upon the value of strip cropping as a soil- and moisture-conservation measure.

Another complaint of biological hue is that burrowing rodents damage conservation structures such as terraces and earth dams. The first plea is usually for artificial control of the gophers, kangaroo rats, or ground squirrels that may be causing the damage. A knowledge of the burrowing habits of these mammals indicates that a broad terrace which can be tilled and worked with farm machinery is not subjected to damage by mammals. A dam possesses a biological factor of safety that prevents failure caused by animal damage if it has sufficient freeboard (height of dam above the spillway) and is thick enough to prevent penetration of burrows into the line of saturation (Compton and Hedges 1943). In California, conservation structures designed well enough to withstand damage from storms were found to be safe from damage by rodents. Thus a knowledge of animal habits can beneficially influence engineering design to the extent that it may eliminate the cost of mammal control and assure the functioning of conservation structures.

PLATE 13

Man's use of the land determines its pattern and its productiveness over the years. To fit farm operations to the land, contour cultivation takes an important and conspicuous place. In Texas, contour strips and terraces cross farm boundaries to make a community pattern (top). In South Carolina the shape of things to come is forecast by alternating strips of cotton and small grain, with a contour-planted orchard in the background (bottom).

[114]

The natural science of conservation practices on agricultural land is also illustrated by the relation of hedgerows to injurious insects. It was once generally thought that vegetated fencerows and woody hedges harbored multitudes of insects harmful to crops grown in adjacent fields. Thus burning of hedges and wayside vegetation was advocated for its supposed insect-control value. But hedges are desirable as fences, windbreaks, snow-moisture retainers, erosion buffer strips, and guides for contour cultivation. Recently, biologists associated with land-management problems have looked closely at hedgerows, and in Ohio have found that very few insects in fencerows are crop pests, most of them are harmless, and a few are beneficial. Sixty times as many aphid-eating lady beetles occurred in osage orange hedges and shrubby fencerows as in bluegrass cover along fences, and hedgerows on many farms studied were the only places where beneficial insects, such as lady beetles, assassin bugs, and damsel bugs, could live undisturbed from year to year. Furthermore, beneficial mammals, as the short-tailed shrew and least shrew, live in fencerows where the woody cover also provides homes for shrikes, sparrow hawks, skunks, and weasels that feed on rodents generally (Dambach 1942). For the control of specific pests, destruction of some types

---

PLATE 14

TOP. A sub-surface tillage instrument whose V-shaped blades are drawn through the soil 2 to 4 inches below the surface. Most used in crop areas of the Great Plains, such implements leave a surface mulch of grain stubble which reduces wind and water erosion, increases moisture penetration, improves soil condition, and controls weeds by cutting off their roots. Grain may then be drilled into the loosened earth with no further soil treatment. Similar implements draw vertical spikes, under-surface blades, or rotating bars through the soil.

BOTTOM. Special power implements strip seeds of native grasses for conservation plantings. Such machines can work over rough ground and harvest seeds rapidly, reaping them before they overripen and shatter. Gramas, dropseeds, and other wild grasses never before commercially available have thus been brought into widespread use. Other implements harvest, treat, and plant seeds and seedlings of both herbaceous and woody species.

[115]

of brushy vegetation is occasionally necessary, but the encouragement of selected perennial plants along fencerows and for hedges is now conceded to be much more of an asset than a liability.

Burning to control insect pests is another practice once condoned that is now less frequently, though still too often, encouraged. When it is known that many beneficial mammals, birds, and even insects live in well-attended, unburned vegetation, it appears profitable to prevent fires. A closer inspection of the lives of some of our insect pests will also show that burning cannot actually destroy them. Grasshoppers pass the winter as eggs within the soil where fire does not reach them, and the corn-ear worm and cotton boll worm also hibernate well below the soil surface. White grubs and wireworms burrow below the frost line. Many harmful insects—army worm, chinch bug, corn borer, and others—resort to burrowing deep into corn or grain stubble where ordinary burning does not harm them (Fenton 1940).

Pasture burning tends to increase weeds like the croton, which, in the southern states, is used by the cotton flea hopper for overwintering. Cotton boll weevils frequently hibernate in trash and litter about cotton fields, but a winter temperature of 10 or 12 degrees above zero kills the hibernating weevils, and so most of them are destroyed in normal winters. Many other troublesome insects spend the winter, when burning is usually done, as eggs or pupae in places where burning cannot affect them. The more we know about their life histories, the more we realize that cultural practices are more likely to reduce harmful insects than will aggressive methods like burning. Furthermore, burning tends to perpetuate annual plants, among which are numbered so many of our troublesome weeds, while protection from fire encourages the growth of more desirable perennials.

That farming involves complex ecological patterns is illustrated in the following example. For a long time agriculturists advised against the use of a permanent ground cover of herbaceous vegetation in irrigated pear and avocado orchards of southern California. It was believed that a ground cover of perennial vegetation sheltered field mice and other rodents, which ate the bark and roots of trees, provided protection to injurious scale insects, increased irrigation costs by absorbing water, and, through competition for nutrients, reduced fruit yields. Orchards maintained for a great many years in permanent cover of herbaceous perennial legumes and grasses, however, have demonstrated that yields are not reduced under this practice. On the contrary, packing-house records show yields from these orchards to be much higher than the average; they also bring better than average prices. Furthermore, operation and irrigation costs are low. Irrigation is accomplished by flooding from furrows constructed across the slope, the water being spread by the dense plant cover which, by providing infiltration into the soil in place of rapid runoff, actually conserves water.

Permanent cover in these orchards obviates annual seeding, practically eliminates cultivation costs, prevents interference with feeder roots of the trees located near the surface of the ground, and reduces soil erosion. The ground cover harbors predacious insects and ground-nesting birds that help naturally to combat insect pests. As an additional source of income, sheep are occasionally pastured on the cover of herbaceous vegetation. This causes no damage to the orchard, in fact helps to protect it against mice. Even if gophers in some soil types must be artificially controlled, the reduced cultivation and protection against erosion far more than pay the cost. Thus the invaluable soil resource is permanently protected, and harmonious relationships are established among the biological components of the orchard community which it is to man's advantage to understand and maintain.

Land-Use Planning

The establishment of conservation practices can only be widespread and effective when it is carefully planned. It may be well at this point to discuss this subject briefly, for a great deal of thought has been given to planning land use in the United States during the past decade. Considerable regional planning has been undertaken by the National Resources Board. Published reports on geographic regions, as the Pacific Northwest, New England, or the St. Louis area, deal with immediate and urgent problems, policies, and organization which can provide for planning, construction, and operation of public works within the region. Conservation and development of forests, irrigation and power projects, flood control, and other public works receive emphasis, although wildlife, conservation laws, preservation of natural areas, soil and land-use surveys, transportation and communication plans, and related subjects may be studied and recommendations made. Such planning is especially useful in co-ordinating state activities that reflect upon regional welfare.

Many kinds of broad planning relate particularly to agricultural activities. Among these are the delineation of problem-area groups, which represent areas where soils, physiography, erosion, climate, and other factors combine to present a condition uniform enough so that a particular program of land management is more or less generally applicable (Finnell 1939). Various states have for some years given attention to planning on a state basis, following land classifications as described in the preceding chapter, and this sort of activity has more recently crystallized through the co-operative effort of farmers and ranchers, the land-grant colleges, other state and local agencies, and the United States Department of Agriculture. One of the results has been the formation of state and county land-use planning committees that attempt to work out plans and policies to co-ordinate agricultural programs, facilitate their function, and develop new programs as needed. Such planning has been undertaken by local groups in nearly one-third of the 3070 counties in the United States, with the number of farmer members looming large.

The medium through which plans and programs of land management are being most effectively applied is the soil conservation district. Such districts are legally constituted subdivisions of a state, set up under a state soil conservation districts law, and in many states their boundaries coincide with those of counties. Established through local referendum and operated under the supervision of a locally selected group of land operators, the district applies its own program of land use in co-operation with any agency, federal, state, or private, that is in a position to provide it with technical or material assistance. The real accomplishments resulting from the work of upwards of 1000 of these districts throughout the country—in actual adjustments and work on the land as well as the planning for such work—augur well for what can be expected from the group, or democratic, approach to meeting problems in the wise use and management of land.

The ultimate success of any land planning is the way in which it applies to an individual operating unit of land. From the ecological standpoint, the excellence of the farm or ranch plan is the criterion for evaluating the success of planning. The individual plan represents the relation between environmental conditions and the operator who lives upon the land, and in so far as possible it should represent an optimum habitat condition that will support the operators for an indefinite period. One of the successful attempts to relate the farmer or rancher to the land conditions upon which he depends is the farm plan developed between the farmer and the soil conservation district to which he belongs, with the assistance of various governmental and other agencies. Such a plan is reproduced in Figures 5, 6, and 7 (pp. 120-21). It is based on the classification of land according to use capabilities.

Rural Zoning

No consideration of land planning is complete without mention of rural zoning. Whenever such zoning is related to capabilities of the land, or some other evaluation of land from the standpoint of its most desirable use, it becomes ecological, and not far removed from Life Zones as conceived by Merriam and other biologists. Most of us are familiar with city zoning, even though we may only recognize it as the obvious pattern of residences here, industrial plants there, and business houses elsewhere. Much of the pattern of the modern city, of course, results from adaptation

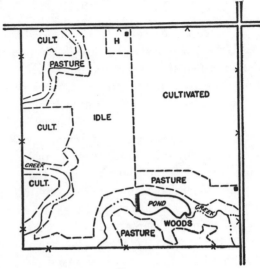

FIG. 5.

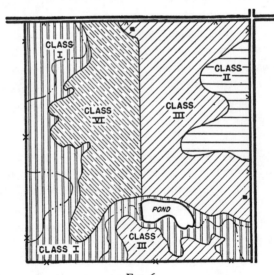

FIG. 6.

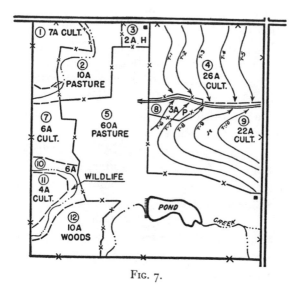

FIG. 7.

FIGS. 5, 6, and 7. The three figures illustrate land planning and management in accordance with land capabilities and adjusted use on an actual 156-acre farm in Wagoner County, northeastern Oklahoma. Figure 5 is a map showing land use before adjustments. Figure 6 shows classes of land according to use capabilities. In Figure 7 encircled numbers indicate fields, the dash-x lines indicate fences. Cultivated fields 4 and 9 are contour tilled and terraced, runoff emptying into a grassed outlet between the fields shown by double line. Fields 2, 5, and 8 are pastures, the last a temporary one. Water is available to both permanent pastures, which extend to the farmstead, Field 3. Cattle are fenced out of the entire west side of the farm, protecting the woods, wildlife area, and the three small cultivated fields. Note that present use is a compromise, as it usually must be, between land capabilities and needs of the operator as shown by past use. The success of the plan, inaugurated in 1938, is proved by reduced runoff and soil loss, increased yields of the cultivated crops, chiefly oats and cotton, ample improved pasture for the dairy herd, locust posts from the planted woods and quail in the wildlife area. (Courtesy of the Soil Conservation Service.)

to physiography or other landscape features. Thus industrial plants are clustered on a river floodplain where access to river and rail transportation facilities is most convenient. The business section is frequently on an adjoining river terrace, while the residential district is scattered at higher elevations beyond.

Out of these early adaptations, however, many of them completely unplanned, city ordinances developed which marked boundaries within which no new stores or factories could be built. In a similar way, rural zoning ordinances designate areas where cultivation of land, or other specified land uses, cannot be undertaken in the future. Most zoning regulations do not interfere with present operations of people who are already located in an area at the time the zoning provisions are adopted. Rural zoning provides a way in which a county, by adopting local ordinances, can guide the location of new settlers and prevent further expansion of undesirable land uses now occurring. In most cases, the county must obtain authority from the state legislature, through an enabling act, before rural zoning restrictions can be enforced.

Much rural zoning is founded upon a land classification adopted for the county or state. It therefore greatly supplements attempts to use the land properly according to its long-time productiveness. Frequently zoning results in developing forests with permanent sources of income by contributing to forest protection and development, and it aids in the dedication of recreation areas most useful to that purpose. It also tends eventually to bring rural people together in more compact communities, closer to roads and schools, where public-health work can be more effective and better public services are available at comparatively low cost. It helps to prevent settlement of land inherently incapable of supporting people who attempt subsistence farming, and although it cannot prevent improper land management by present operators, it contributes much toward eventual readjustment compatible with environmental conditions.

Although it may seem odd to think of new settlement in a country as well occupied as ours, there are always shifting, expansion, and conversions in use. In the five years from 1930 to 1935, for example, nearly 600,000 new farms were reported established in the United States, some of them put to the plow for the first time. Even considering the fact that a fair number of these farms had been in existence but were not recorded previously, the figure emphasizes the changes in land use which constantly occur. Census figures usually show about 6,000,000 farms in operation in the entire country in any one year.

Rural zoning can be most effective when combined with such other land-use adjustments as suggested by individual farm plans. Furthermore, a public land purchase program for a zoned area can enable stranded and isolated farm families to move to more desirable locations. Several states now have land purchase programs and the Federal Government is also engaged in some work of this type.

FORESTS

LAND USE AND FOREST TYPES

RECORDS of primeval forests in the United States are few—indeed one searches almost in vain through the literature of early American travelers for satisfactory descriptions of them. The forest types, as the forester calls them, which today compose the forested areas of our country are largely 'second growth.' There are a few small patches of comparatively undisturbed woodland scattered in eastern United States, and some large tracts in the western mountains that essentially represent climax types, although the latter are almost everywhere disturbed by grazing livestock. For the most part, our present forests represent stages in succession, and should be looked upon not as permanent landscape features, but as plant communities that are changing and likely to be quite different in the course of a few decades.

Such an attitude greatly influences one's thinking with respect to forest management. For example, instead of planting nursery-grown white pine promiscuously throughout New England because white pine occurs there in pure stands, the modern forester recognizes that those stands in most places indicate a stage in old-field succession certain to be replaced eventually by trees of a different kind—in this case a climax of mixed hardwoods and conifers. The story of forest succession in mid-New England has been pieced together so beautifully by foresters at Yale and Harvard Universities that it is given here in résumé. (Plates 15, 16, 17, and 18, pp. 126-7, 130-31.)

The primeval vegetation of central New England consisted of a mixture of coniferous and deciduous trees (Harvard College 1941). Many species occurred, for the forest was in reality a broad ecotone or transition type. To the north spread a forest of hard maple, beech, yellow and paper birch, basswood, poplar, red spruce, hemlock, white pine, Norway pine, and balsam fir. Trees of this forest type or community were usually found in the more northern, protected areas of central New England. Trees characteristic of the more southerly forest, on the other hand, were found in this region on more open exposures—trees such as red, white, black, and scarlet oaks, hickory, chestnut, tupelo or black gum, black birch, and pitch pine. Intermediate sites were characterized by a mixture of species from the two forests, plus several others including white ash, black cherry, and red maple. Pure stands of white pine originally formed a permanent type only on light, sandy soils, and a transient type on burns or blowdowns.

Settlers began to move westward into central Massachusetts about 1700, and for half a century small clearings were cut and burned in the forest depths to provide tillable fields for an expanding population. By 1830 the landscape had totally changed, for in place of forest-covered hills, verdant green with an unbroken blanket of interlacing boughs, there stretched cultivated acres as far as the eye could see. Up to 80 per cent of the area was in tilled fields, orchards, and pasture, and the remaining scattered woodlots were grazed and culled for wood products much needed on the thriving farms.

But a change was to take place—a transformation as startling as the clearing of the land. Beginning shortly after 1830 and continuing for several decades, this region witnessed a period of farm abandonment. Richer farmland in the West, railroads, new industrial centers, gold in California, and the Civil War all helped to drain the New England hills of the men whose ancestors had so recently wrested them from the wilderness. Almost before the people had gone, nature began to reclaim the fields. Following the invasion of weedy plants, white pine seedlings sprung up everywhere, as did a scattering of hardwoods. The conifers grew so fast they suppressed the other trees, and solid stands of pine resulted. So extensive were these old-field stands that between 1895 and 1925 fifteen billion board feet of second-growth white pine were cut in central New England, with a manufactured value of more than $400,000,000. Clear cutting was the rule, for there was as little management to the lumbering as there had been to the settlement 200 years earlier.

Then an interesting thing happened—only explicable if one understands something of plant succession. The cut-over pine stands did not come back to pine. Instead, young hardwoods, many planted by rodents and birds, which had started to grow in the shade of the pines and which had been cut out of the way by the lumbermen, began to sprout. Most of them were red maple, red and white oak, white ash, hard maple, chestnut, black cherry, and black birch. Sprouts of such origin make good timber trees, although sprout clumps from the large stumps of hardwoods that grow up with the pine are worthless for saw timber. If adjacent white pines are seeding, which they do every other year, some pines may pioneer among the hardwood sprouts, of course, and during the first few years after logging, seedlings of other light-demanding species like gray and paper birch, pin cherry, and poplar also appear. A similar process would take place if, instead of lumbering, the pines had been destroyed by windfall or fire, as happened under primitive conditions.

About twenty years after logging the pine, many of the desirable timber trees were overtopped by inferior stump sprouts and the fast-growing weed trees such as birch, cherry, and poplar. Such a stand, if undisturbed, would produce trees useful for cordwood and low-grade knotty lumber only. If they were cut for cordwood, another stump sprout generation of inferior trees would

---

PLATE 15

TOP. Originally the primeval forest of central New England stood with great trees of white pine, hemlock, beech, sugar maple, white oak, yellow birch, hickory, and many other noble species.

BOTTOM. By 1730 much of the virgin forest was felled, and here and there early settlers removed stumps for home and crop sites. As in all early agriculture, the cultivated fields rather than the pastures were the areas fenced. (See Plate 16.)

appear. This succession following logging of pine, so evident during the past few decades, has convinced many persons that nursery planting of more desirable species was the only recourse for productive forestry in much of New England.

The successional behavior of these trees, however, teaches another lesson. If, before a pure stand of pines is clear-cut, the seedlings and small sapling undergrowth are cut off at the ground, they will, after the pines are removed, develop into straight, thrifty trees. Six or seven years after cutting, it is necessary to thin the stand which develops in order to remove the inferior species, the several-stemmed sprouts from occasional stumps of large hardwoods in the pine stand, and the weed trees. Such thinning works with the natural process, and is much less expensive than the use of nursery-grown stock.

On poor, light soils where white pine may originally have occurred as a soil or edaphic climax, pine can be favored by partial cutting, thus creating a favorable seed bed on the forest floor and supplying seedling pioneers with sufficient room and light. When the new trees coming into this cut-over area are safely rooted, another group of adjacent pines can be felled, and so the rotation made to proceed over the years.

The succession which occurs on cut-over lands is frequently different from that which occurs on old fields. After lumbering, in many eastern hardwood areas, oaks and frequently other species sprout from the old stumps. Here, also, management becomes a matter of thinning to favor the species most desired.

In Washington and Oregon, when the climax forest of western

PLATE 16

TOP. A century later nearly 80 per cent of the land was cleared and in some sort of agricultural use. The forest remnants were grazed and culled for wood products needed on the thriving farms.

BOTTOM. By 1850 rich land of the West and new industrial centers had drawn many farmers from their land. Brush grew over the stone walls, and in the old fields the seeds of white pine found a good place to take root. (See Plate 17, p. 130.)

hemlock, western red cedar, and other conifers, is destroyed, succession occurs which in principle is comparable to that in eastern forests. Logging is almost invariably followed by fire, after which fireweeds (*Epilobium* and *Erechtites*) and other herbs appear, to be followed frequently by blackberries, bracken fern, and shrubs such as California hazelnut and *Holodiscus*. Eventually these plants are succeeded by a stand of Douglas fir, which is ecologically comparable to the stands of white pine in the Northeast. Douglas fir also starts to grow on raw mineral soil immediately following fire and it forms solid stands on abandoned fields in the Northwest. The Douglas fir in time gives way to hemlock, red cedar, and other original conifers, which eventually form an uneven-aged stand of near-climax composition. This process may be interrupted by repeated fires, when a heavy cover of bracken fern develops to persist as long as the area is regularly burned. When burning ceases, succession again proceeds, although hemlocks may then get a head start by their capacity to grow under the shade of the brackens.

A knowledge of natural revegetation emphasizes the fact that forest plantations of a single species, row upon endless row, are no more tenable from a successional standpoint than they are from the standpoint of maintaining soil fertility or insurance against insect and fungus attack. Even the foresters of Europe, where once the tidy stands of single species were regarded as the ultimate in forest management, have of recent years directed their thinking toward the 'Dauerwald' doctrine, which recommends selective cutting from a mixed stand simulating a natural forest.

The one-stand forests, established a century ago in Germany, were a bold and useful step in revegetating exhausted and deteriorated land, but they probably would have served their purpose more successfully if they had been consciously established as an artificial stage in a succession directed toward a forest that in composition was essentially a climax type. Heske has dealt with this whole problem in his book on *German Forestry* (1938), in which he writes:

German experience confirms the biologic fact that the forest is a complicated community of living beings, in which each tree species is merely a member, no more and no less important for the health of the whole than the other members. A single species may not be used with impunity in plantations where it is entirely isolated from its natural organic complex. The foundation and the elements of practical silviculture are not the individual species of trees, but the natural life communities of which these economically desirable tree species are a part. The growing of commercially less valuable, but biologically important, species in mixture with those of high economic value is equivalent to paying an insurance premium against later losses.

The abundant German experience with these monocultures is already of value as a lesson for the rest of the world. The lesson will become even more convincing in the future, as it becomes possible to demonstrate in specific instances what caused the frequent failures of the monocultures. These may have been due to the pure stands as such, to the failure to adapt species to site, to the even-aged, schematic form of the forest, to other factors, or to the joint effect of several or all of these. This important problem has by no means been solved. In place of theories, we must have more experiments before a conclusive judgment is possible.

As factors influencing forest types, fire and grazing are very important. Fire may not only destroy trees, it determines in many instances the type of forest that follows. The fact that many trees and forest communities may be used as indicators of burns proves the extraordinary effect of fire. There is good reason to suppose, as already mentioned, that the present composition of much of our southern pine woods is due to repeated burning, and that without fire a forest predominantly deciduous would succeed the pines. The blueberry-covered hills of the northern Appalachians are maintained by fire where hardwood or mixed forests would otherwise occur, and many other types of vegetation owe their existence to fires, even though the burning may have occurred many years ago. Extensive lodgepole pine forests in the Rocky Mountains, for instance, can be traced to fires antedating the establishment of the pines.

As noted below, grazing can be an influence of considerable

effect upon woodlands. Wherever a forest is subjected to heavy grazing, its future as a reasonably permanent plant community is jeopardized. Not only may tree seedlings be destroyed, but even shrubby and herbaceous elements of the flora may be materially changed. Whether grazing is by an abnormally high number of big-game animals in an extensive forest or by milk cows in a farm woodlot, the effect may be equally damaging. It is now generally conceded that a farm woodlot cannot be economically productive of wood products and livestock at the same time, and that large forests must maintain carefully controlled numbers of big game and cattle if the forest is to support such animals and be permanently productive of timber (Plate 19, p. 138). The land-management biologist must be alert not only to existing conditions which are the result of fire and grazing in the forest, but fully cognizant of the ultimate results of such use of forest lands.

FORESTS AND ANIMALS

American ecologists have given a great deal of attention to forest animals and their effect upon the welfare of the forest. A few instances of such relationships are cited below.

BIG GAME

In the western states, the big-game problem is well illustrated by the now famous example of the Kaibab deer. In the Kaibab Forest, on the plateau north of the Grand Canyon in northern Arizona, a national game preserve was established in 1906. Deer

---

PLATE 17

TOP. By 1910 old fields were well covered by stands of pine, which reached merchantable size in 60 to 70 years. Trees of all sizes and ages were cut indiscriminately, with no provision for another crop.

BOTTOM. On the land just cleared of pine, logging operations were not followed by another growth of pine. Instead, sprouts came from the stumps of hardwood undergrowth, originating from seed stored in the duff by animals or blown in from near-by stands after logging. (See Plate 18, top.)

shooting was prohibited and government hunters were hired to kill mountain lions and coyotes. As a result, mule deer increased until they had practically exterminated all palatable shrubby vegetation and had browsed, as high as they could reach them, the junipers and pinyons, which, with yellow pine, form the dominant vegetation over much of the plateau. Furthermore, cattle were permitted to graze in the forest, thus competing with the deer for food. By 1920, the range was taxed beyond its carrying capacity and deer died of starvation. It was not long before it became apparent that something would have to be done, for not only was the existing vegetation devastated, but all tree seedlings providing woodland reproduction were essentially eliminated. In order to restore a semblance of balance, predators were protected, live-stock grazing was more carefully controlled, and hunting regulations were made less restrictive.

Thus a paramount problem of the forest ecologist has come to be reduction of numbers of wild animals, as deer and elk, to the point where these mammals do not have a damaging effect upon the forest vegetation. One of the reasons for the present abundance of deer, both in eastern and western forests, and the great numbers of elk in the wooded mountains of the central Rockies has been the protection afforded by laws designed to conserve them. Purely natural factors, however, have been just as important in determining numbers of big game within our forests.

PLATE 18

TOP. Today young hardwoods form a well-stocked stand, largely of fast-growing weed trees. By careful thinning of the inferior species, sprouts of the more desirable timber trees can be encouraged. Thus a forest much like the original one, in balance with natural conditions, can be reproduced and eventually maintained with a minimum of cost. (This series consists of photographs of dioramas in the Harvard Forest Museum, Petersham, Mass.) BOTTOM. In many parts of the country where terrain is flat and winds strong, windbreak plantings are used to protect crops and hold the soil from blowing. Here rows of tall eucalyptus trees protect California citrus groves from damaging winds.

[131]

The forest succession in central New England, as outlined in the first part of this chapter, gives a clue to the situation in the eastern states. When the original forests or one of their successional stages, as a stand of white pine, were lumbered, hardwood sprouts appeared, which provided excellent deer food. This food, made generally available throughout the Northeast not long after the turn of the century, together with laws permitting the taking of bucks only, created an ideal condition for the deer, which increased until farmers cried for protection from the animals that moved from neighboring woods into orchards and crop fields. Just as surely as this second-growth browse matures into forest trees, however, the deer will diminish in numbers, for food will become scarce and many deer may starve. Game managers see this eventuality, and already are experimenting with the cutting of scattered blocks in maturing hardwood stands to learn how to provide sufficient food for a desirable deer population.

Realizing that various kinds of timber cuttings greatly influence the forest as wildlife habitat, the Society of American Foresters' Committee on Game Management recently issued the following statement (Chase et al. 1942):

It is a common misconception that a forest of a particular type is unchanging, so that it has a constant potentiality for wildlife production. The fact is that its potentiality varies with age, density of stocking, treatment, and area. Thus, in New England, a young, even-aged white pine stand at the time it first closes is used for cover by cottontails, ruffed grouse, and other species, but its food production is practically nil. When the boles have a couple of feet of dead length, the cottontails no longer use anything except the borders where live limbs reach the ground. By the time the stand reaches an age of about 25 years, competition has reduced the number of trees enough to enable the first, most tolerant herbs to appear. As the stand approaches maturity, enough of the subdominant trees have died out to allow considerable development of herbaceous plants, shrubs, and advance growth of tree species. At 60 years of age, food production reaches its peak. If the stand is held longer, advance-growth hardwoods develop a closed canopy of their own, again forcing out the herbaceous food plants (Gould 1935). Obviously, then, the time of cutting in such a stand largely determines what food and cover species for wildlife are to follow.

Until sportsmen and the general public realize some of the ecological links between big game and the forest, there is likely to remain objection to laws or regulations that permit the taking of female deer and elk or the slaughter of surplus animals, even for the meat they might provide. No range should be stocked beyond a carrying capacity compatible with its maintenance in a permanently productive condition. When the habitat as influenced by man encourages an unnaturally high population of animals injurious to man or to a condition which he wishes to support, in this case the forest, such population must be reduced, even if purely artificial means are necessary as emergency measures to effectuate such reduction.

SMALL MAMMALS AND BIRDS

Literature accounts much damage to forest trees by mammals. Porcupines, squirrels, rabbits, mice, and other animals have been accused and persecuted because they eat bark, injure young seedlings, or destroy tree seeds. Frequently the obvious relationship between these animals and the forest is the one upon which action is taken. But, as Grinnell (1924) has pointed out, there is another side to the story, and 'the forest trees themselves depend, for their maintenance in the condition in which we observe them in this age of the world, upon the activities, severally and combined, of the animals which inhabit them now, and which have inhabited them in the past.' It has also been noted that forest birds have a distinct insect-destroying value, and that the pocket gophers, ground squirrels, moles, and badgers are natural cultivators of the soil. Furthermore, many animals, as woodpeckers, nuthatches, chickadees, tree squirrels, chipmunks, porcupines, burrowing beetles, termites, and ants, contribute importantly to the comminution of vegetable matter and thus hasten its return to productive soil.

[133]

That mammals are not all bad was early recognized by those in America close to the use of the land, with whose observations and ideas most of us are all too unfamiliar. In a letter to Richard Peters, dated 18 December 1809, Dr. B. S. Barton quaintly wrote:

. . . Almost everyone believes, that our mole, which I have no doubt infests or visits your ground, for it is very common along the Schuylkill,—that the common mole of Pennsylvania, is a very pernicious animal. I wish you could turn the attention of some of the members of the agricultural society to this subject. It is one of no small consequence. I greatly doubt if this mole be so pernicious as is imagined. I have long entertained doubts on the subject . . . By loosening the earth, and thereby enabling the radicles of different plants to progress with more facility; and by devouring a portion of the radicles which it meets with, does not the mole of the United States, do quite as much good as harm? . . . May we not, by preserving moles from unnecessary destruction, turn their fur to useful purposes in the United States?

Forest rodents serve a highly useful purpose in planting seeds, a service frequently underestimated by foresters themselves. The inauguration of higher stages in plant succession, as the invasion of pine stands by hardwoods, results in no small part from the planting of heavy seeds by small mammals. Plants bearing heavy seeds like acorns and nuts, whose dispersal might depend otherwise on gravity, have their seeds carried uphill and distributed widely by jays, squirrels, and other animals. It has been pointed out that black walnuts and other nut-bearing trees in California owe their distribution almost entirely to rodents and that the stands of Douglas fir which follow logging and fire in the Northwest spring largely from seeds cached in the ground by rodents, sometimes resulting in as many as 40,000 seedlings per acre (Hofmann 1923). In the northern Rockies, mice, which are good planters because they cache individual seeds rather than cones, establish western yellow pine on grassy slopes where fallen seeds do not easily grow. The invasion of black oaks on the sand plains of south-central Connecticut, along the Quinnipiac River, is due more to squirrels which cache the acorns than to any other influence. Caches of acorns are found as far as 200 feet from the nearest mature oak (Olmstead 1937). Direct seeding of forest trees by man has not proved very successful, and rodents have been blamed, among other things, for this failure. This is not surprising in view of the fact that rodents are forever busy about their business, which deals so much with seeds, including the eating of them. On the other hand, management of forests on a naturalistic basis cannot afford to ignore the service these small mammals render.

Another very interesting role of rodents in forest succession has recently been described. On the rocky, lava plateaus of the Pacific Northwest, the seeds of higher plants cannot find foothold from which to grow. In the soil beneath the broken lava, however, pocket gophers work and, in the normal course of their activities, work some of this soil up to the surface of the rocks. There it serves as a place for plants to germinate. On these patches of soil, grass grows, and eventually the yellow pines of that region take hold. The establishment of a forest cover on these rocky expanses is thus made possible by the work of a small burrowing mammal (Dalquest and Scheffer 1942; Larrison 1942).

Although some forest animals may actually be injurious to the forest, as deer in large numbers, it is almost invariably more profitable to try to correlate management operations with the behavior of animals than to fight them blindly. A good example of working with nature, not against her, is that concerning the snowshoe hare of the Lake States forests, often charged with intolerable injury to young trees, especially on clean-cut or burned-over areas where natural reproduction produces very thick stands. The snowshoe hare is a highly cyclic species, with 'highs' of large populations occurring at 10-year intervals. When the hare population is at its peak, the animals eat, girdle, or prune the young trees until the stand is so open that they may be seen easily by predatory mammals, owls, and hawks. The hares must then retreat to thicker stands for protection. They may reinvade the young trees at intervals of a few years whenever the trees have again thickened enough to form protective cover, and the hares may thin out the stand recurrently until the bark becomes too thick to be palatable. Instead of being an unmitigated evil, however, the opening of the stand permits the remaining trees to recover from their stunted condition, helps to reduce the fire hazard, and minimizes insect damage (Cox 1938).

In northern Minnesota, the value of the thinning operations of the snowshoe hare is set at a high figure. Furthermore, foresters have learned that in this region wherever tree seedlings are planted as a reforestation effort, plantations show a much higher percentage of survival if the trees are widely spaced, in which case they do not provide escape cover under which the snowshoe hares can work, and that there is minimum damage if plantings are correlated with 'lows' in the population cycle of the hare, a species whose numbers fluctuate regularly.

The land-management ecologist must be prepared to analyze a situation correctly when he meets it in the field. In eastern New Mexico, cattle men recently contended that pinyon trees were being killed by porcupines. Although stockmen sometimes wish to destroy pinyons and junipers because they think the trees reduce the grass, in this instance the trees were desired to provide shade for cattle and to protect them from winter winds. The trees were scattered over a long, rolling ridge in heavily grazed range dominated by weedy perennials, chiefly snake weed (Gutierrezia). Examination showed that the trees were well scarred by porcupines, nearly all of them with at least one broad, barkless mark on the trunk or limbs. Many had been heavily chewed, and a number were dead, with brown needles identifying them even at a distance.

There was no doubt that the porcupines, denning in rocky outcrops and escarpments near by, had been at the pinyons. It was evident, however, that the dead trees seemed to occur more or less in groups, and were not promiscuously scattered through the stand. Closer inspection showed that neither the trunk nor the limbs of very many of the dead trees had been ringed by the porcupines, and that no fungus infection seemed to occur on the scars. Further investigation proved that both larvae and adults of a little *Ips* beetle, so destructive to many tree species, occurred in abundance under the bark of the affected pinyons, and their work was evident on the dead trees. The damage thus resulted not from the apparent injury, but from a source at first wholly unsuspected.

Production of wood products is, of course, the primary objective of the management of land adapted to forests, but forests may yield other products also. They can serve as recreational areas, produce wild fruits, and support useful wildlife at the same time that they afford a profitable crop of timber. Wildlife production is illustrated by the methods employed to encourage ruffed grouse, a much desired game bird, in the forests of the Northeast. Such methods involve (1) eliminating fire and domestic livestock, thus protecting cover and food for the grouse; (2) using a selective cutting system, which provides the uneven-aged stand most favorable to the bird; (3) keeping woodland roads brushed out, thereby providing dispersed open areas; and (4) developing food-bearing shrub margins at woodland openings and edges (Edminster 1942). In addition, a forest that supports grouse and other woodland animals is in a naturalistic condition that permits far less insect damage and depletion of soil nutrients than occur in stands of single tree species. It is a useful commentary on old-style forestry that, in Europe, solid stands of pine were subjected to insect damage so severe that hawks and owls were liberated to reduce rodents and artificial nest boxes were hung to attract birds as insect destroyers!

MAMMAL-INSECT RELATIONSHIPS

Although small forest animals, such as shrews, moles, mice, and chipmunks, have been so often listed on the debit side of the forester's ledger, recent studies show that small animals inhabiting the forest eat an astonishing number of insects, many of them larval forms of species highly destructive to mature trees. Hamilton and Cook (1940) found that in forests of the Northeast there is an average of about 100 small mammals per acre, with some woodlands supporting as many as 300 per acre. The proportion of insects in the diet of forest rodents—mice, chipmunks, and flying squirrels—may not be more than 20 per cent, but for the insectivores—shrews and moles—the insect food runs from 50 to 75 per cent. The insect-destroying value of these mammals may be even higher than the value of insectivorous birds, for the number of such animals per acre is much greater than the number of birds, and, unlike most birds, the mammals are resident and usually active throughout the year. Furthermore, they are voracious creatures, many of them eating food equalling nearly one-third of their weight daily. It is conceivable that, without forest animals, we might have very poor forests indeed.

In the forests of Michigan and Minnesota, mice destroy so many larch sawflies, highly injurious to larch trees there, that the number of sawfly cocoons opened by mice can be used as a census method for determining mouse abundance. Inasmuch as the mice consume only the soft parts of the cocoons, the insects are not identifiable in stomach contents, but the incisor marks on the cocoons identify those opened by mice. Of 481 cocoons at Itasca Park, Minnesota (Graham 1929):

74 (15%) emerged normally	15 (3%) were parasitized by insects
382 (80%) were opened by mice	10 (2%) were destroyed by fungi

---

PLATE 19

TOP. One way animals affect plants is strikingly shown by the browse or 'Plimsoll' line caused by sheep and goats on these aspen trees in New Mexico. Deer and other big-game animals cause similar results and in many forested areas have been known literally to eat themselves out of house and home, then starving to death.

BOTTOM. To the left this Ohio sugar maple woodlot has been protected from fire and grazing for 10 years. There soil micro-organisms are three times as numerous as in the grazed woods to the right, small mammals are more abundant, there is tree reproduction, and maple syrup yields are twice as great as in the unprotected area.

This example shows a high percentage destroyed by mice. Studies indicate an average destruction of about 60 percent, with a greater number of mice and consequent cocoon damage somewhat higher in dry areas or dry years.

Similar beneficial effects of mammals have been noted in the West. The pandora moth (*Coloradia pandora*) usually attacks ponderosa pine and Jeffrey pine; it normally injures lodgepole pine only when it occurs with the preceding species. An infestation of epidemic proportion in lodgepole pine forests, however, occurred in north-central Colorado in 1937. The role of predators as an influence on moth numbers was indicated by the fact 'that squirrels and bears were destroying many of the pupae. The bears had overturned many flat stones and the squirrels had dug cone-shaped holes in the ground in search of pupae. Animal feces composed almost entirely of pandora moth eggs were found' (Wygant 1941).

The trees need not be in forests to be subject to the effects of mammal-insect relationships. The most destructive insect enemy of the peach is the larva of a moth called the peachtree borer. Much attention has been given to its control, and animals play a part in this work. Jumping spiders, chrysopid larvae, pigs, moles, and skunks prey upon the borer larvae. According to a recent study in Georgia (Snapp and Thomson 1943) the greatest help comes from field mice and rats, the most important predators of the peachtree borer in that State. These mammals destroyed hundreds of pupae in almost every peach orchard there in 1932, and in many orchards they destroyed about two-thirds of the indi-

---

PLATE 20

TOP. A badly used and severely damaged landscape, of little use, in West Virginia, all too characteristic of many long intensively cultivated areas.
BOTTOM. Planted to black locust and other useful woodlot trees, and protected from fire and grazing, the same spot four years later becomes a potentially productive area.

viduals that pupated in the fall of 1933. The field mice and rats dig the peachtree borer cocoons from around the base of the trees and eat out the contents. Although the borer can be easily controlled by placing a ring of paradichlorobenzene crystals on cleared ground around the trunk of the tree, the value of mammals, birds, and other animals as normal checks on populations of injurious species should not be underestimated. Without such predators, the need for and the cost of artificial control might well be much greater.

FARM WOODLANDS AND WILDLIFE

In planning the best use of his land, the farmer frequently finds some sizable areas too steep or potentially unproductive for crops or pasture. Such land can profitably be maintained in trees, for a good farm forest is the source of useful wood products (Plate 20). Like the larger forest, a woodlot can also yield other products, such as a crop of wildlife. In order to hold soil and maintain natural reproduction of the trees, grazing and fire are excluded from the well-managed farm woodland. This, in itself, is of importance to wildlife, for it makes possible much more shelter and food than exist in a grazed, burned stand of trees.

Studies in Ohio have shown that ungrazed woodland supports more than twice as many pairs of breeding birds as comparable grazed woods (Dambach and Good 1940). Strikingly analogous results were obtained by similar studies in Ontario (Mayall 1938). Both birds and mammals made greatly increased use of a 120-acre weedy tract in central California only three years after it had been planted to trees and shrubs (Hawbecker 1940a).

Detailed comparisons of grazed and ungrazed farm woods reveal amazing differences. A study of grazed and ungrazed woodland in New York (Chandler 1940) has shown measurable differences in climatic factors, such as air temperature, light, and relative humidity. An investigation in northeastern Ohio (Dambach 1944) showed that small animal organisms—millipedes, centipedes, spiders, and insects—which work in the humus and leaf litter of the

[140]

woodland floor, helping to convert it into soil, were about three times as numerous in the protected, ungrazed woods, while small mammals, such as short-tailed shrews, white-footed mice, and pine mice, were one and one-half times as abundant there. All of these organisms aid in increasing the moisture-holding capacity of the soil, reflected in a slight increase in basal area, volume, and number of trees in the ungrazed area, protected for a period of ten years. There was no reproduction in the pastured woods, which was immediately adjacent to the ungrazed plot. (Plate 19, p. 138.)

The most remarkable difference in the two areas, which were sugar maple woodlots, was the fact that the protected woods had yielded an annual production of 21.9 gallons of maple syrup at the rate of 2.2 square feet of basal area per gallon, while the grazed woods produced an annual average of only 16.2 gallons, and required 3.3 square feet basal area per gallon.

The preservation of den trees and an occasional piling of slash from cuttings may make the woodlot more suitable to wildlife. Conservation practices, such as selective cutting for sustained yield, are beneficial, for they maintain a comparatively open and stable woodland habitat. A management measure of interest to the forester and biologist alike is the development of the woodland edge, for in this ecotone between the trees and open field or pasture wildlife is more abundant than within the woodland itself. Many more species of birds, for example, as well as more individual birds, are usually present in the marginal areas.

A modern farm woodlot has a border of shrubs, planted or developed by cutting out the tree species, to simulate a natural woodland edge. Such a border not only provides food and cover for wildlife, but hinders desiccating winds from reaching the forest interior and checks windthrows by deflecting wind above the trees or reducing its velocity. European foresters employ such a border as a 'stormprotector' for the forest, and recommend conifers to reduce the damage occasioned by destructive winds.

Other woody areas on the well-managed farm include hedgerows and windbreaks. Every quail hunter knows to look for birds along old rail fences overgrown with shrubs, briers, and vines. Many farmers believe that overgrown fence-rows harbor harmful insects and are a source of weed seeds; as a result, they have become devotees of 'clean' farming. The fence-row or hedge that consists of a planned and controlled stand of selected plant species, however, can be a useful land-management device. It is now recognized that a living fence or hedge, if planted on the contour, acts as a barrier to soil erosion, and it can include shrub species producing berries and fleshy fruits useful in the making of pies, jellies, jams, and preserves. Certainly hedgerows, tracing a diverse pattern across a countryside, mean a great deal in the way of travel lanes, food, and nesting cover for wildlife, and they help greatly in creating, on the farm, an environment pleasing to man.

Just as hedgerows in the East may help to hold soil when they are established across the slope, so in many parts of the West windbreaks and buffer strips act as effective barriers in reducing wind action and, in northern states, retaining snowfall (Plate 18). Whether in the form of a narrow windbreak of a few rows of trees flanked with shrubs, or a buffer strip of a single row of shrubs, such plantings may perform the dual role of controlling wind erosion and creating on sub-humid agricultural land, where woody plants are scarce, a habitat suitable to many game and insectivorous birds.

THE range country stretches from the short-grass plains through the forested Cordillera and Great Basin to the desert grasslands of the Southwest, California, and the Cascades of Oregon. Few Americans realize that the western range occupies nearly 40 per cent of the land surface of the United States. It is a vast area and one of our most important land resources. Many of us think of the range country as little changed by man. Yet in how many parts of the short-grass plains, just east of the Rockies, can we now find conditions as Casteñada, traveling with Coronado into the High Plains, described them 400 years ago (Winship 1896):

Who could believe that 1000 horses and 500 of our cows and more than 5000 rams and ewes and more than 1500 friendly Indians and servants, in traveling over those plains, would leave no more trace where they had passed than if nothing had been there —nothing—so that it was necessary to make piles of bones and cow dung now and then, so that the rear guard could follow the army. The grass never failed to become erect after it had been trodden down, and although it was short, it was as fresh and straight as before.

This was not of the lush prairie that is now the corn belt, but of the kind of land that bred the Dust Bowl!

Original types of range differed with climate and other environmental conditions, and were almost as varied as the kinds of forest. It is beyond the scope of this book to consider vegetation types, however, except as they relate to conditions influenced by man. Consequently, as with forest types, range types are treated only with reference to land use.

LAND USE AND RANGE TYPES

As with existing forests, the types of range in the United States today represent far less the original communities of plants than they represent kinds of vegetation resulting from past and present use of the land (Plate 21). Each range or forest type, of course, is related to the climax vegetation once occupying an area, but what we see today is usually a successional stage determined largely by man's influence. To be able to recognize range types for what they suggest in the way of past conditions and possible improvement is the first task of the range manager—whether technician or operator. Range management involves a great deal more than manipulation of plant succession, but the range manager who does not recognize the significance of changes in vegetation is handicapped from the start. (Plate 21, p. 146.)

With a knowledge of succession, one can tell from the plant cover of a disturbed area not only how far existing vegetation is removed from the original plant cover, but also, if he observes it over a period of years, whether the vegetation is progressing toward the climax or retrogressing toward a lower stage in succession. Although not much has been done to apply such knowledge, a generation ago ecologists learned that, in the wheatgrass grazing land of central Utah, the species of plants tell one of two stories (Sampson 1919). If the new plants appearing on the range belong to a stage lower in succession than the predominant vegetation, the range is deteriorating, a condition brought about by misuse, as overgrazing. If invading plants belong to a higher successional stage, then the range is improving.

The situation may be diagrammed:

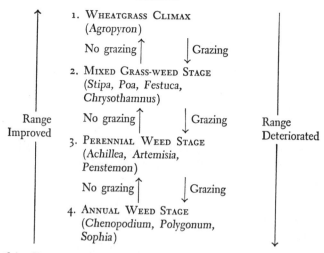

1. WHEATGRASS CLIMAX
   (*Agropyron*)

   No grazing ↑   ↓ Grazing

2. MIXED GRASS-WEED STAGE
   (*Stipa, Poa, Festuca,*
   Chrysothamnus)

   No grazing ↑   ↓ Grazing

3. PERENNIAL WEED STAGE
   (*Achillea, Artemisia,*
   Penstemon)

   No grazing ↑   ↓ Grazing

4. ANNUAL WEED STAGE
   (*Chenopodium, Polygonum,*
   Sophia)

Range Improved

Range Deteriorated

In this diagram the wheatgrass stage at the top represents the original vegetation. Of course, many other kinds of herbaceous plants, both grasses and forbs (non-grasslike herbs), occurred along with the wheatgrass, but it was the prevailing dominant. After cattle were introduced and had grazed this grassland for several years, the wheatgrass became sparse and grasses less common in the original vegetation began to spread, giving rise to stage 2. A few perennial weeds also showed up more prominently as the original grasses disappeared and, with continued grazing, most of the grasses fell out of the picture and the perennial forbs, common and undesirable enough to be termed weeds, began to predominate. They indicate grazing heavier than desirable and are represented in our diagram as stage 3. The fourth stage, composed of annual weeds, is characteristic of range so badly overgrazed that even the perennial weeds do not grow there. When cattle are taken off the range, or sufficiently reduced in number, this trend tends to reverse itself, stage 4 passing into stage 3, and so on. Because the variety of palatable plants makes the second or mixed grass-weed stage the most desirable for grazing all classes of stock, the range manager will attempt, by controlling the number of livestock, to maintain the range in this condition.

The species of plants in the different stages may vary from place to place, especially between areas originally occupied by distinct types of grassland. The vegetation type, that is, annual weed, perennial weed, and so on, represented by each stage, however, remains essentially comparable in different regions. There is, on the whole, greater variation between the higher stages from one place to another than there is between the lower or earlier ones. The plants of the annual weed stage, in fact, are remarkably the same throughout the range country, witness the widespread occurrence of Russian thistle, cheat grass, and many other common range weeds. Even some perennial weeds, as snake weed or turpentine weed (*Gutierrezia*) and sneezeweed (*Helenium*), are also well distributed.

It is of interest to note that no amount of money spent on weed control when the range is in the annual weed stage, or even the perennial weed stage, can successfully eliminate weeds. The only economic way to control range weeds is to remove the cause of their appearance, which is so frequently overgrazing. If grazing pressure is relieved, and some of the higher-stage grasses persist in places to provide seed, the natural process of plant succession will inevitably replace the weeds with more desirable plant species —and at no cost to man. It is not at all unusual to read, in discussion of a particularly troublesome weed that 'its increase in

---

PLATE 21

TOP. The contrast between overused and well managed range is graphically shown on adjacent pastures in New Mexico. The grass is thicker and constitutes a plant community of distinct species on the carefully managed area to the left. There are also fewer weeds and troublesome rodents where the grass is better.
BOTTOM. To the left, where grazing was heavy, the grass cover on this Colorado range was so light that the snow readily melted. To the right the grass holds the snow, runoff is retarded, and moisture infiltration is increased. Environmental effects of land use are shown by different vegetation and micro-climates on the two sides of the fence.

abundance on most areas of range no doubt has resulted in a marked and serious reduction in grazing capacity.' But we should put first things first, and realize that range weeds—snake weed (*Gutierrezia*), prickly poppy (*Argemone*), sneezeweed (*Helenium*), horsebrush (*Tetradymia*), burroweed (*Aplopappus*), and many others—do not cause poor range, but are themselves the result and indicators of intensive foraging by grazing animals.

If it seems to the reader that the ecologist, with his fine theories, can tell the stockman nothing of any practical importance which the latter has not learned by hard experience, it might be well to cite the rancher who, after 30 years of running cattle on the same range, was influenced to reduce his herd in accordance with the recommendation of a range ecologist. In the course of a couple of years, the rancher for the first time saw a number of range grasses put on recognizable growth and produce seed. His interesting remark was, 'I always knew there was grass in there, but it never got big enough for me to know what kind it was.'

It has been contended, particularly by some geologists, that the deteriorated condition of our range land, especially as it is expressed by erosion in the Southwest, is not the result of overgrazing by domestic livestock, but is due rather to a general change in land level which would have caused widespread soil removal and streams to cut arroyos whether man and his stock were present or not. There are others, chiefly stockmen who do not wish to have anything interfere with their running as many cattle as they choose, who contend that the present deteriorated condition of the range in most parts of the West is due to lack of rainfall, and 'if and when it rains,' the depauperate range vegetation will flourish again as lush as ever.

PLATE 22

A valley in Navajo County, Arizona, photographed about fifty years ago. The stream flowed in a shallow bed, and water from it was readily available for irrigating small tilled fields. (Compare Plate 23, p. 162.)

The land manager will not deny the geologist nor the cattle-men their arguments, but the poor grass, the range weeds, the eroded cattle paths, and the deep arroyos are there for all to see and for some to live with (Plates 22 and 23). And if the prevailing vegetation represents a stage low in the successional scale, it is necessary, even in the face of periodic prolonged drought and geological complications, to bring the range up to a desirable stage where it will support a reasonable number of cattle over a long period of time. Much can be accomplished toward this end, if, in accordance with land-use capability classes or other guide, the necessary reduction in stocking and other appropriate management measures are applied. (Plate 22, p. 147; Plate 23, p. 162.)

ANIMALS AND THE RANGE

The range manager assumes the task of maintaining an optimum population of domesticated animals on a natural stand of grass, and at the same time relating this operation to populations of wild animals that are an integral part of the natural plant-animal community composing the range.

PREDATORS

On range land, predation has been frequently looked upon as a problem that could best be solved by destruction of the predator. Because a wolf, coyote, mountain lion, or bear is known to have killed a lamb or calf, we are prone to war promiscuously against all of their kind. It is difficult to say how much money has been spent on predator control in the United States, but the sum of public funds expended by the Federal Government in co-operation with the states, counties, and livestock associations for predator and rodent control in the fiscal year ending 30 June 1941 amounted to nearly three million dollars.

Perhaps it is not so disgraceful that we fanatically try to kill every range rodent and pursue the last predatory mammal to its den as it is short-sighted to spend scarcely a penny on the study of animal populations that would teach us more of the true be-

havior of wild creatures and more intelligent ways of dealing with
them. Dr. Lee R. Dice, who ranks as one of America's foremost
mammalogists, stated our negligence of facing fundamentals of
mammal control when he wrote (1938b):

. . . When naturalists advocate that scientific studies be made
before beginning a control campaign, they are told there is no
time to make a scientific study while a farmer's crops are being
eaten up or his poultry devastated. This excuse would be more
satisfying if it were not repeated year after year. For over twenty
years, in my memory, I have heard the same story, with only slight
variations. The ecologic and economic studies of wildlife which
should be made in every region before initiating a poison campaign are, to the best of my knowledge, not now being made, and
with a few notable exceptions, never have been made anywhere
[in the United States] by men competent to conduct such studies.

The much maligned, though sly and artful, coyote is the object
of attempted annihilation, although it has increased in numbers
and distribution during the years of its persecution. At almost any
large gathering of ranch operators one is sure to hear vituperative,
profane epithets heaped upon this animal. Such invective must
have its reason, but this kind of approach, which unfortunately
permeates a good deal that is written about the coyote as well as
much that is spoken, is scarcely the way to solve an ecological
problem.

What does the coyote eat, as a matter of fact? A recent study
on the ecology of the coyote, designated by the Wildlife Society
to be the outstanding paper of the year (Murie 1940), summarizes
the food of the coyote in Yellowstone National Park, based on
analyses of more than 5000 coyote droppings. Almost 65 per cent
of the food proved to be rodents—field mice, gophers, hares, mar-
mots, etc. Less than 20 per cent consisted of big game, of which
more than half was elk, domestic cattle being practically non-
existent in the Park. In 1941 there also appeared an excellent study
by Charles C. Sperry, which analyzed the contents of more than
8000 stomachs collected throughout 17 western states during the

five-year period 1931-5. Sperry showed 76 per cent of the coyote's food to be carrion, rodents, and rabbits. Only 13½ per cent consisted of sheep and goats, some of which was undoubtedly carrion. It is largely because of this portion of the coyote's food that he is condemned, for comparatively few coyotes eat calves, colts, or pigs. The remaining 10.5 per cent of the coyote's food consists of deer, birds, insects, other animals, and vegetable matter.

Because those administering the Tule Lake Wildfowl Refuge in northeastern California wished to protect waterfowl from all harm, it was suggested that bobcats and coyotes be 'controlled' on the immediately adjacent Lava Beds National Monument, where predators were not intentionally disturbed by man. To determine the effect of the coyote on its prey, R. M. Bond (1939) made a careful study of this relationship, based on three years' collections of 273 scats and 9 coyote stomachs. Although the Monument supported a large number of sheep, only 12, composing about 0.1 per cent, were killed by coyotes during a one-year period. This in spite of the fact that 10,500 sheep grazed the area during three months before the lambing season where the coyote population was about 2 per section. The simple and inexpensive expedient of using road flares to keep the coyotes at a distance from the flock at night was largely responsible for the low sheep loss.

Incidentally, these studies showed that insects may form a considerable part of the coyote's diet, 20 per cent or more in some instances, which indicates that even large animals may serve to influence insect numbers. It is also of more than passing interest to learn that the coyote has been proved recently to be decidedly resistant to botulism of the type that causes such widespread devastation among waterfowl. Consequently biologists are now looking with relief upon the coyotes in the waterfowl refuges, for they make a habit of patrolling lake shores to eat ducks that are very sick or dead of the disease.

In western United States, a method of sheep management that succeeds without killing coyotes is to keep the sheep in large paddocks of 1 to 5 sections of land surrounded by a coyote-proof fence.

Another involves a string of coyote-proof corrals distributed over the range. In some cases this even permits the sheepherder to go back to the ranch every night to eat and sleep, returning in the morning to herd the sheep so as to reach another corral by night-fall.

This brings us to the economics of predator control, a matter seldom well aired. Although the cost of killing a coyote in some parts of California is reported to be many times as much as that in northeastern California where, according to officers of the Tule Lake Refuge, it is $1.40, even the latter conservative figure may not be commensurate with the need for control. As Bond points out:

Even a few coyotes might wreak havoc among sheep that strayed away or were neglected by the herders. Only a very large (and hence temporary) overpopulation of coyotes would greatly endanger sheep that were properly cared for. It thus seems that even if they could control coyotes for $1.40 each, the local sheepmen could hardly justify, on an economic basis, any very heavy control program, even off the Monument, if they had to pay for it themselves instead of getting it done 'free' by the government.

Very few persons will contend that control is entirely unnecessary, any more than they would refuse to spray the beetles that damage the beans in their gardens. But land-management decisions should be based upon thorough consideration of the interrelationships involved. If a coyote is observed to kill a lamb, a fact is established, namely, that coyotes kill lambs. Since lambs are desirable, the obvious conclusion is that coyotes must be destroyed if lambs are to survive. Of rabbits and rodents that live throughout the range of the coyote, another isolated fact is apparent—rabbits and rodents eat grass. This simple relationship seems also to interfere with an activity of man, for the cattle and sheep he herds likewise eat grass. The obvious conclusion is that rabbits and rodents should be destroyed to preserve food for livestock. These are the simplest deductions—but they do not represent ecological thinking.

The biologist wants to know how the destruction of coyotes affects other things. Although we can say that coyotes live on rabbits and various small rodents to a substantially greater extent than on lambs, the practical biologist wants to know how many lambs are killed by coyotes and how many coyotes would have to be killed to save them. Also, what is the cost of killing the coyotes as balanced against the value of the lambs saved. Finally, how do the coyotes affect other animals that compete with the lambs for forage?

It is not easy to learn all these interrelationships, and we need to guard against the common sort of reasoning that assumes a direct cause-and-effect relationship between phenomena that coincide in time or space. Events that coincide in time are not necessarily related, although many correlations have little more than such coincidence to support them. There may be a seven-year sunspot cycle and a corresponding seven-year variation in the level of Lake Titicaca, but that doesn't necessarily mean that sun-spots cause the lake level to vary. It is reminiscent of the story about the German village with the highest birth rate in the region. There were also more storks nesting on the roof tops than in any other village. Hence the relationship between the stork and the baby.

RODENTS

If any one controversy about rangeland has raged more vehemently than that over the advisability of reducing predators, it is that of rodent control. Many studies upon which control is founded are themselves based upon what happens on small observational plots that are scarcely indicative of vegetative conditions on large acreages. Furthermore, the behavior of wild mammals confined in small plots cannot be expected to exemplify the behavior of those animals over vast acreages. Until we can compare range conditions over large areas—thousands of acres in extent—as a watershed where certain rodents are present against a comparable watershed where they are absent, and measure the result in pounds of beef or mutton produced, we are likely to fall short of a knowledge of the practical relationship of rodents to range vegetation.

It is of advantage to know something of the life histories of rodents, and here again funds spent on population studies could reap dividends. Kangaroo rats are frequently accused of destroying range vegetation, and methods for their destruction are freely advised. At first glance this may seem logical, for certainly kangaroo rats eat range plants—live on them entirely, in fact. It is important to know, however, that rodents of many kinds seem to prefer annual grasses and succulent weeds to perennial grasses. Recent studies show that the food of kangaroo rats consists predominantly of annual plants (Hawbecker 1940b; Monson 1943; Monson and Kessler 1940). Monson claims, for example, that banner-tailed kangaroo rats 'are not found common in country dominated by perennial grasses, which do not provide an abundant and unfailing crop of seeds as do the annual grasses and weeds.' If kangaroo rats and many other range rodents are dependent to such an extent upon annuals, they could scarcely be so abundant in a higher successional stage composed primarily of perennial vegetation. Here is an indication of the fact that many range rodents are 'animal weeds' and that their abundance can be related to the occurrence of weedy plants.

It is generally observed, also, that jack rabbits are usually most numerous where the range is poorest and the grass most sparse (Piemeisel 1938). Even prairie dogs, whose occurrence is not known so definitely to coincide with range type, may often belong with the 'animal weeds.' Black-tailed prairie dogs are not known to have lived in the shinnery savannah of western Oklahoma when it consisted of a good stand of tall grasses dotted with low, scattered clumps or 'mottes' of dwarf oaks. When this land was broken into farms and livestock grazing reduced the original grasses, changing the stand to plants more like those of the mixed prairie, in patches of which prairie-dog colonies originally occurred in this region, the prairie dogs began to invade the changed shinnery vegetation. Destruction of predators and reduced food supply within the original boundaries of the colonies accentuated the movement. The animals have now become so abundant that they heavily browse both the oak and the sand sagebrush (Artemisia filifolia) that were natural, woody components of the original vegetation. Here, then, is one more example of a rodent thriving in a habitat of which it was not originally a part, but which has been made suitable to it through changes brought about by man's use of the land (Osborn 1942).

In general terms, it may be stated that most rodent populations are comparatively light on climax grasslands or rangeland in a high successional stage. Conversely, rodents are relatively abundant in vegetation types characteristic of land long overgrazed or of abandoned cultivated fields, where the cover is composed of non-climax annual grasses, other annual herbaceous plants, or shrubs such as sagebrush. There are a few rodents, like meadow mice, which may have a different relationship to successional trends. These tend to decrease with overgrazing, largely perhaps because they like the protection and nesting materials provided by the grasses. Also, because they are diurnal, they are very vulnerable in short herbage to crows, hawks, and other predators. On the whole, changes in land management, which cause a desirable change in plant cover, look once more like the solution to a problem so closely allied to use of the land.

Studies of rodents that attempt to show a relationship between rodent populations and vegetation at a particular time must fully consider, in addition to the effects of the animals, current climatic or other conditions influencing the vegetation. That variations in the composition of vegetation may be marked from one year to another has been nicely pointed out by a recent work (Talbot et al. 1939) that showed great fluctuations in stands of annual vegetation in California. Based on a number of observations in the southern San Joaquin Valley, an interesting comparison was disclosed between the vegetation in 1934, when growing conditions were poor, and that in 1935, when they were good. In 1935 twice as many species of plants occurred with a single species of the composite family, unrecorded in 1934, occupying 13 per cent of the herbaceous cover. While 3 species formed 97 per cent of the vegetation in 1934, 15 species constituted this proportion of the plant cover in 1935.

Other students (Weaver and Albertson 1939) have also shown that major changes in grassland can result from continued drought. In the central United States during 1934-7, even dominant perennial grasses were so greatly reduced that the composition of the vegetation was materially altered. Such wide fluctuations in the composition of the plant community are directly related to range management because of the changes in forage they effect, and to soil conservation because of variations in protective ground cover that result. They emphasize, also, the danger of tabulating effects of rodents upon vegetation except over a long period of years when all factors can be given a reasonably weighted average.

The land manager will want to acquaint himself with the various kinds of rodents, and to become familiar with their habits. All too often we hear of rodent damage or rodent control as if a rodent was a particular animal or all rodents were alike. The relation of rodents to land differs with the specific habits and behavior of the species concerned, which vary considerably from beaver to groundhog to muskrat to mouse. Among range rodents, as the prairie dogs, ground squirrels, kangaroo rats, and pocket gophers, there are many distinctive traits, and the mammalogists remind us that the rabbit is sufficiently different from all the others to be kept apart as a lagomorph, or special type of gnawing mammal. Of ground squirrels alone the biologist recognizes nearly 100 different forms in some 30 distinct species; similarly there are many kinds of each of the other main rodent groups.

Within a particular group of rodents, such as the prairie dogs, for instance, species may vary in habits as they do in appearance and geographic distribution. The black-tailed prairie dog of the open grasslands of the Great Plains, for example, builds a rather conical mound of upturned earth which is kept bare of all vegetation within a considerable distance from the burrow. Furthermore these dogs are highly colonial. By contrast, the Gunnison prairie dog of the Southwest, where shrubs are usually conspicuous components of the vegetation, although living in towns, is much less gregarious than the black-tailed species; its mounds are broad, flat-topped, and more widely distributed, and the surrounding areas are permitted to support a cover of vegetation often denser than the prevailing plant cover, the result probably, of fertilization by the animal's excrement, cultivation of the soil by which they increase its moisture holding capacity, and removal of weeds. It would appear, therefore, that the black-tailed prairie dog not only does more apparent damage to vegetation, but could more easily be exterminated by poisoning, since it keeps to well-defined 'dog towns.' There is also a difference in the food, for the black-tailed animal has a diet of which more than 60 per cent is grasses, while the Gunnison prairie dog draws less than half of its food from plants of the grass family (Kelso 1939). Such differences in the habits of animals with which the land manager must deal emphasize the necessity for discriminating among even closely related species.

On much western range land, depleted by intensive use, and on dry farm lands abandoned and in an early successional stage, it is frequently considered desirable to plant grass seed to hasten establishment of useful vegetation (Plate 24, p. 163). Frequently climax or sub-climax native species, such as western wheatgrass, are used. Some exotics, as crested wheatgrass, are favorites for this purpose because they readily produce a luxuriant growth. There is considerable question about the need for reducing rodents on lands being reseeded, but since rodents are frequently most numerous on land needing revegetation, control may well be required until a desirable stand of grass is established. Although rodent control is a costly operation too frequently unrelated to benefits obtained, it is more economical as an initial adjunct to reseeding programs than when it is conducted year after year as the primary effort to improve conditions on badly depleted range.

Range reseeding, if species used and methods employed are soundly selected, may well be a desirable practice under some conditions, as most pasture seeding is useful. Effort should be made to relate it closely to expected changes due to natural successional development, however, and range managers agree that it is usually more practical and economical to obtain recovery by more natural means. Where 15 per cent or more of the soil surface is covered with perennial grasses, the range is likely to recover by careful livestock management alone. Systems of deferred grazing to permit the grasses to store food and produce seeds, furrowing, or the rotation of pastures so that each receives regular protection, aid greatly in the restoration of depleted range areas.

INSECTS

The land-management biologist needs a working knowledge of the taxonomy, habitat requirements, and life histories of plants and animals, especially those that occur in common features of the landscape, such as crop fields, pastures, woodlands, and interspersed hedges, streambanks, and roadsides. The fact that certain leaf hoppers live on range weeds in the western United States and northwestern Mexico is in itself not a very useful fact. Neither is it particularly significant that the hoppers in the spring migrate to beets, tomatoes, and other cultivated crops, for they do little damage by eating those crops. When it is learned, however, that one of these leaf hoppers carries and transmits the virus of the very destructive disease known as curly top of sugar beets and western yellow blight of tomatoes, a knowledge of its life history becomes highly important.

Interestingly enough, the host plants that support the leaf hoppers through the winter are nearly all annual weeds occurring on overgrazed range land. Most important of the summer and fall

[157]

hosts is Russian thistle, a plant most abundant on abandoned farm lands and poor range. The plants that support the beet leaf hopper from one beet crop to the next, therefore, represent an early stage in plant succession occurring on lands that have been poorly managed, that is, depleted range or crop fields no longer tilled because they were inherently unproductive. Management methods that will change the vegetation to a stage higher in the successional scale will automatically remove most of the intermediate hosts of this troublesome pest, and it is highly interesting that one of the most effective methods recommended for the control of the beet leaf hopper is the elimination of overgrazing in order to increase the more desirable range grasses.

Even the abundance of grasshoppers, so destructive of crops and range vegetation, can be related, in some degree, to intensity of land use. Weese (1939) has been credited with the statement that a barbed wire fence is the best device for controlling range insects, meaning that controlled grazing which permits good grass will automatically reduce the number of injurious forms. His statement was based largely on his own investigations, one of which is summarized in Table 4.

TABLE 4. RELATIVE NUMBERS OF RANGE INSECTS, WICHITA MOUNTAINS, OKLAHOMA

| | NUMBERS OF INSECTS | | RATIO OF INSECTS IN OVER- |
| | Overgrazed (Short grass) | Normal (Little bluestem) | GRAZED TO NORMAL |
INSECTS	Range	Range	GRASSLAND
Beetles (Coleoptera)	23	10	2.3
Flies (Diptera)	5.6	5.5	1
Bugs (Hemiptera)	12	2	6
Cicadas (Homoptera)	43	10	4.3
Bees and Wasps (Hymenoptera) ...	28	6	4.6
Grasshoppers (Orthoptera)	36	4	9

Under the conditions of this study, the overgrazed area supported about four times as many individual insects as the normal grassland. The normal grassland, however, contained more species of insects than the overgrazed range; the former harbored 90 species, the latter 72. The greater abundance of grasshoppers in the overgrazed area is believed to result from: (1) more succulent young grass shoots serving as food, and (2) more suitable areas for oviposition, for most grasshoppers prefer to lay their eggs in bare ground.

Soil conservationists concerned with the prevention of soil blowing in the northern Great Plains find that on fields in which the grain has been harvested with a combine, leaving the straw stubble on the ground, or on fields covered with straw mulch as a wind erosion control measure, there is comparatively little grasshopper oviposition.

Grazing Land and Wildlife

The proper management of range land is of general benefit to desirable wildlife because it permits the development of a good grass stand that creates shelter, food, and water for many wild birds and mammals. Antelope, sage hen, prairie chicken, and other natural animals of the Great Plains and the intermontane West can better be conserved by maintaining an environment as natural as proper range use will permit. As already pointed out, there is reason to believe that jack rabbits and rodents, as 'animal weeds,' are less abundant on well-vegetated range. There is a definite relation between wildlife welfare and conservation practices on range land, and studies by Gale Monson (1941) indicate that the number of birds on moderately grazed range is approximately twice that on comparable overgrazed areas. The development of well-distributed stock ponds, a sound range-management practice, may be as beneficial in providing water to wildlife as to livestock.

The intimate relation between range conservation and soil conservation has long been emphasized by far-sighted persons, and wildlife benefits are known to accrue through the establishment of conservation practices. On semi-arid range lands of the western states, water-spreading and water-retention devices, such as contour ridges, furrows, or chiseling, that is, fissures cut in the soil by a narrow steel blade, produce a heavier stand of grass. It is now believed that contour furrows to facilitate water percolation in grasslands are most effective if they consist of comparatively small grooves, about five inches broad and five inches deep, which do not disturb the soil greatly. Such furrows, of course, are of little use in heavy soils in which they readily fill, or in sand which does not hold firm. Turning over large furrows or constructing ditches frequently creates disturbed areas populated by annual plants that serve as a source of weed seeds. Experiments are required to determine for each area the type of furrow that will provide maximum water percolation and minimum disturbance of the grass cover.

At first thought it may seem that grazing land does not matter much to wildlife. When ranges are in a good stand of grass, however, or pastures are maintained in grass-legume mixtures designed for sustained forage yield and maximum erosion control, they may supply much in the way of both food and nesting cover for wildlife. Rotating pasture systems, using two or more kinds of pasture to provide a maximum of good forage throughout the grazing season, may provide undisturbed areas for ground-nesting birds. Where deferred grazing is practiced as a forage-conservation measure, attention to nesting dates can provide time enough for birds to raise a brood before grazing is permitted. Meadows of permanent hay, frequently established as part of a strip cropping system or otherwise as part of a well-planned farm layout, provide benefits to wildlife similar to those furnished by well-managed pastures.

There is another significant wildlife benefit to be derived from the establishment of pastures as part of a balanced conservation farm plan. By converting eroding areas that are too steep for the safe cultivation of clean-tilled crops into pasture protected by a cover of grasses and legumes, livestock grazing may be limited to areas more specifically adapted to that use. Thus stock may be kept out of fields and woodlots that are, as a result, rendered more suitable wildlife habitats.

WILDLIFE

THERE was a time when man hunted as much as he liked, and so he still does in some parts of the world. Although many kinds of animals once seemed so abundant that they could never be exterminated, it became apparent long ago that we could not kill wild creatures promiscuously if we were to keep them with us. Prohibitions were then enforced, of which the earliest is probably one of the oft-quoted laws of Moses:

If a bird's nest chance to be before thee in the way in any tree, or on the ground, whether they be young ones, or eggs, and the dam sitting upon the young, or upon the eggs, thou shalt not take the dam with the young:
But thou shalt in any wise let the dam go, and take the young to thee; that it may be well with thee, and that thou mayest prolong thy days. (Deuteronomy 22:6, 7.)

Until rather recently, restrictions against killing them were believed sufficient to perpetuate wild creatures. When desirable types of native animals became scarce, in spite of prohibitions intended to preserve them, people turned to substitutions. Exotic species, introduced even in ancient times because of the curiosity and excitement they aroused, came to be considered easy replacements for native birds and mammals no longer abundant.

The history of wildlife management in the United States recapitulates experience with wild animals in other parts of the world. Connecticut was the first state to regulate game seasons and prohibit export, in 1677. Iowa in 1878 first limited the amount of game to be killed, and Wisconsin in 1887 prohibited the sale of protected game. A long list of substitutions had its beginning with the introduction of the Hungarian partridge into New Jersey in the year 1790. An early idea in game management that still has its advocates was predator control, which assumed that desirable animals could be preserved by exterminating their enemies.

In time, we began to look to the establishment of public works and the setting aside of lands especially devoted to wildlife for the perpetuation of desired species. The first wildlife refuge created by law in the United States was set up in 1870 in California, followed by Indiana in 1903, and then Pennsylvania, where an extended refuge system was started in 1905 (Gabrielson 1943). These, together with the establishment in 1903 of Pelican Island, off the coast of Florida, as the first Federal wildlife refuge, inaugurated a series of Federal, state, and private game refuges that have since spread throughout the nation.

The next movement was centered around scientifically sound management, together with controlled harvest of a consciously produced surplus. With these was combined education to inform the people who harvest wildlife of the work and objectives involved. Sound management recognizes that an environment has a saturation point, and that a habitat can support only a given number of individuals of a particular species. Even under ideal conditions, numbers are restricted to a maximum per unit of area. Food, cruising radius, territory, and other factors make it impossible to support for long a density of wild animals greater than that which such environmental conditions permit. The best we can do is to establish an environment as ideal as we know how for a species that we wish to encourage, and depend largely upon that environment to produce the desired results. Recognizing this condition,

---

PLATE 23

Today the stream flows through a deep channel. The buildings were purposely removed. Deepened arroyos, eroded range, weeds, and poor grass all indicate undesirable conditions which the land manager is called upon to remedy. Compare Plate 22, p. 147.

pen-raised animals are now usually released only after they have first been carefully conditioned and where the habitat has been made suitable for them. Primary importance is attached to the development of natural populations.

The idea of habitat improvement as a basic practice in game management is more and more becoming a dominant one. William R. Van Dersal (1940) stated this idea when he wrote:

The tradition and the law that wild animals belong to the State have for a long time deluded many into believing that wildlife has no connection with the land. In its extreme form this idea has led to the releasing of game animals on land hopelessly barren of food or cover. It has also resulted in the philosophy that the production of game for hunting amounts almost to a duty on the part of the landowner. Not so very long ago, however, an idea began to filter into wildlife literature that has had a profound effect upon wildlife management practice. This idea, even yet not commonly accepted, embodies the conception that wildlife is a product of the land, and that as patterns of land-use change, so do wildlife populations.

After pointing out that a large proportion of our country is agricultural land, and that agricultural land managers hold the key to much of the wildlife that will be produced in the future, Van Dersal emphasizes his point by listing strip-cropping, farm ponds, hedges, field borders, and other conservation measures as good land-use practices that will support wildlife. Such types of improvement can be integrated so closely with land management that there is no conflict between this type of wildlife production and

---

PLATE 24

TOP. Heavily overgrazed, this New Mexico range area provides extremely poor forage for livestock and supports grasses so sparse that they are no longer able to hold the soil.
BOTTOM. Four years later natural plant succession, seeding of adapted grasses, and proper handling of livestock have checked erosion and developed a more desirable vegetative cover.

[163]

other agricultural production. As long as wildlife conservation can be so combined with other land uses as to make the practice of it necessary for reasons other than wildlife production, then whether or not wildlife is of immediate interest to the landowner makes little difference—wildlife is still produced.

The period into which we seem now to be moving, then, is one in which all of the useful ideas embodied in what has gone before, especially the management phases, are combined. In addition, it sees the further step of integrating the management of wildlife with the management of all types of land. Such integration is dependent upon and facilitated by the classification of land and will make possible a more desirable production of wildlife, especially since it results from action programs now being conducted by those who operate the land, with assistance from properly qualified technical sources.

One of the foremost students of general biology in the United States, W. L. McAtee, whose many interests have led him far in both scholastic and applied spheres, has said that 'the livest, the most widespread, and perhaps the most socially significant activity in the field of American biology today is the technology known as wildlife management' (1936). Those interested in wildlife management now form a large group, represented by a professional organization known as The Wildlife Society, which publishes a quarterly journal dealing with various technical aspects of the work of its members. The Society numbers among its constituents foresters, soil conservationists, range managers, biologists, and others, all trained or experienced in the management of wild plants or animals, most of them concerned with the manipulation of wildlife as it relates to the use of some type of land.

Wildlife and Natural Principles

The successful wildlife manager is an arch ecologist, and it is not possible in this chapter to cite more than a few examples of the natural principles of particular concern to this technician.

Many other principles already mentioned in previous chapters are employed by him.

Just as we came to learn that restrictions or prohibitions upon the taking of game could not alone provide the abundant fishing, hunting, and outdoor recreation which men in a free world like to look upon as their inalienable right, so we also learned that we could not 'substitute' an exotic bird or mammal for one we had thoughtlessly caused to become scarce or absent, or successfully depend upon 'stocking' an area with animals bred in captivity. Game managers now look carefully at predator control, attempting to evaluate predators as natural components to be considered as part of a complex environment. Today even the idea of setting aside inviolate areas of public lands—the sanctuary or refuge—is accepted as an accessory or stopgap means of supporting populations of desirable wildlife, unless the management of those areas is carefully integrated with appropriate use of adjacent lands.

The old idea that a refuge served as an area of propagation—a natural game farm—was encouraged by a number of irrelevant facts. For example, it was once believed that the increase in deer in Pennsylvania resulted from the system of refuges in that State, although it eventually became plain that the abundance of second-growth browse throughout the Commonwealth was the primary cause of their large numbers. There is now good reason to believe, as Edminster (1937) has pointed out in his analysis of the value of refuges for cyclic game species, that, although 'the possible function of a refuge for preserving seed stock on areas that have an *abnormally* high hunting pressure, as those near large cities, has not been tested, . . . it would seem . . . that such areas would only function to prevent extermination, not as a means of furnishing continued hunting on surrounding lands.'

Edminster's survey of ruffed grouse on the Pharsalia Game Refuge and on an adjacent area open to hunting in southern New York during the three years 1935-7, showed that for two of the three years the refuge carried fewer birds than the hunted area. The refuge comprised 2120 acres, the check area 1893 acres. In

1935 the refuge had only 66 per cent as many grouse as the hunted area; in 1936 the populations were almost the same on the two areas; and in 1937 the refuge had about 70 per cent as many birds as the hunted area. Numbers of cottontail rabbits and gray squirrels also indicated no consistently higher populations in the refuge than on the hunted area. As the ecologist will learn to recognize, proper habitat, whether we call it a refuge or not, serves to support a surplus of game which, if it does not succumb to disease, the weather, or other natural influences, may be taken by man as a result of regulated hunting without material effect upon the population of that game.

Like the forester, range ecologist, and agriculturist, the wildlife manager can make great use of the idea of succession. It tells him the direction in which vegetation is progressing; it indicates past land use and suggests future trends in landscape development; and it serves as a guide for his recommendations, as when and where to suggest the planting of herbs, shrubs, or trees for wildlife food and cover, and the species to use. With the possible exception of the range manager, the wildlife manager has probably tried to manipulate succession more than any other land technician. By disking, burning, and flooding, he has attempted to encourage and control plant communities favorable to his purpose.

That man, by proper management, can maintain native vegetation in a condition suitable to his needs, that he can 'arrest' succession, as already suggested, is an idea valuable in range and pasture management, forestry, and nearly every land-management practice dealing with plants. The wildlife manager was quick to utilize this principle, as exemplified by the work of Herbert L. Stoddard in the Southeast. Stoddard has demonstrated how to maintain, beneath open stands of longleaf pine, a heavy undergrowth of native leguminous plants valuable as food for the bobwhite quail. This is done by carefully controlled, light periodic burning of the forest floor. Such a practice also aids in maintaining the forest of longleaf pine, for burning encourages reproduction of this species, and the resulting heavy herbaceous vegetation may well provide better grazing, although little attention has been given by ecologists to these further relationships. The course of plant succession and the result of burning as a quail-management practice in southern Georgia may be diagrammed as follows:

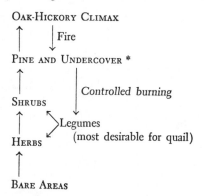

OAK-HICKORY CLIMAX

Fire

PINE AND UNDERCOVER *

Controlled burning

SHRUBS

Legumes
(most desirable for quail)

HERBS

BARE AREAS

* The stage commonly dominant for the 'Southern Pine' region, due to recurrent, uncontrolled, heavy burning, even prior to the advent of white men.

Some other facts of interest in connection with this practice are worth mentioning. The 'brown spot' disease which attacks young longleaf pines is believed to result from a deficiency of calcium and phosphorus in the soil. These elements are apparently made more readily available by burning, for the disease is less damaging then than when fire is excluded. Although the longleaf may grow more slowly as a result of the burning, it consequently produces more valuable timber. In this region, where the average rainfall is more than 50 inches, recorded precipitation one year fell to 32 inches. As a result many big pines in non-burned areas died of drought, due to competition from the deciduous trees that were invading the stand, while the pines in burned areas were not subjected to such competition and so survived.

The significance of the ecotone or 'edge' is fully appreciated by the accomplished game technician. Although this subject has been considered in Chapter II, the effect of edge on bird populations is illustrated by the following example. In Walker County, Texas, a study of the birds in contiguous types of vegetation illustrates the fact that the margins support more species and higher numbers of individual birds than the interior of the same vegetation types. The results of the study (Lay 1938) are shown in Table 5.

TABLE 5. EDGE EFFECT DISPLAYED BY BIRDS, WALKER COUNTY, TEXAS

	MARGIN			INTERIOR		
TYPE OF VEGETATION	No. Counts	Avg. No. Species	Avg. No. Birds	No. Counts	Avg. No. Species	Avg. No. Birds
Oak-palmetto river bottom	3	6.0	17	2	3.5	4.5
Oak-elm river bottom	3	6.7	19	4	4.25	6.25
1-4 yr. cutover pine	2	6.5	15.5	3	5.3	13.3
10-14 yr. cutover pine	3	3.3	10.3	5	4.4	8.2
15 yr. and older cutover pine ...	3	9.7	20.7	2	6.0	11.0
Summary— all counts	14	6.5	16.6	16	4.6	8.5

The successful treatment of a transition area or ecotone on agricultural land is exemplified by the management of the border between a crop field and adjacent woodland. In the Gulf and Atlantic Coast States, there has long existed an erosion problem perplexing alike to the farmer, forester, and agronomist. The problem relates to the edge of crop fields adjoining woodland, where the shade and root competition from the trees prevents or stunts the growth of the crop. Usually, for a distance of 20 to 30 feet from the edge of the woods, the crop in the field grows poorly; frequently it fails completely in the rows closest to the trees. On such areas, water also accumulates from the crop rows and, coursing down the edge of the field, washes the soil and may create a useless and unsightly gully.

The biologists interested in soil conservation saw in this area an opportunity not only to save soil but also to develop a strip of vegetation useful for wildlife on land otherwise valueless (Plate

25, p. 178). In order to create desirable cover on such a site, some plant was needed that was at once a herbaceous perennial, able to grow in a depleted soil, and capable of competing with trees for sunlight and soil moisture. A satisfactory plant was found—the introduced Asiatic *Lespedeza sericea*. As now established on thousands of farms in the Southeast, border strips of this plant effectively prevent erosion, afford turn rows for teams working the field, and furnish a useful habitat for wildlife (Davison 1939). In order to produce a more satisfactory edge to the woodland and to provide a more beneficial cover and food supply for wildlife thus encouraged, shrubs are frequently developed between the herbaceous strip and the trees. On such sites in the South, introduced shrubby lespedezas (*L. bicolor* and *L. cyrtobotrya*) can be successfully and inexpensively established by direct seeding. Both there and in other regions trials with native shrubs show that some of them also can be grown in field plantings from seed.

The establishment of field border plantings has been eagerly adopted by the farmer and has been found equally acceptable to the agronomist and forester. Biologists have observed cottontail rabbits, bobwhites, and other birds in the border strips, and studies to determine the effect of such strips upon populations of insects, birds, and mammals are under way. Bicolor lespedeza has already proved to be a favorite food of the bobwhite quail. It is also the source of a light, mild honey. As a honey plant, it is noteworthy, for it flowers in late summer when few other honey plants are in bloom. Evidence is accumulating to show that field borders support insectivorous bird populations several times as numerous as those of near-by meadowlands, depending somewhat on the amount of woody cover in the border.

WILDLIFE LAND

It has long been acknowledged that the best use of extensive expanses of marsh land is the production of muskrats and waterfowl, and it is easily conceded that rough mountain and canyon country, in addition to its scenic value, is a good place for wild creatures. The idea that there are many smaller parcels of land scattered throughout the country better suited to the production of wildlife than to the growth of any other crop is a rather new one. Perhaps the first to point out the potential value of these areas was Aldo Leopold, who stated in his game survey of the North Central States (1931 p. 249) that 'if the cover needed for watershed conservation were restored to the drainage channels and hillsides of the North Central region, the upland game problem would be half solved.' Later Stoddard (1932 p. 5), Leopold (1933), and Darling (1934) further emphasized the desirability of converting eroded areas into wildlife habitats. In his suggestions for pheasant management in Michigan, Wight (1933) pointed out that the revegetation of eroded gullies improved wildlife habitat, and a year later Grange and McAtee, considering the improvement of the farm environment for wildlife, treated the value of thickets and other types of wildlife cover for checking erosion. About this time, Ernest G. Holt (1934) began to stress the importance of integrating wildlife conservation with soil conservation, and during the succeeding years won consideration for wildlife welfare in the rapidly developing nationwide program of soil conservation aimed by the Federal Government at better use of agricultural lands. Wildlife management is now accepted as an integral part of those programs that involve conservation and adjustment of farm and ranch lands. That management measures beneficial to wildlife are most economically undertaken when they are a part of a co-ordinated land-use program has been attested by an economic study of wildlife as a supplementary farm enterprise, undertaken jointly by the United States Fish and Wildlife Service and the Bureau of Agricultural Economics (Miller and Powell 1942). The investigators concluded that the most profitable objective of wildlife management on agricultural land is a 'co-ordinated conservation program that will integrate wildlife production and utilization into all land-use and soil-conservation programs.'

It is of the utmost importance that, in accomplishing his purpose, the wildlife ecologist think first of the major use of land, then of the plants or animals he wishes to manage. He cannot expect permanently to succeed with methods that *superimpose* habitat improvements upon methods for producing crops, livestock, or timber. Instead, his methods must become a part of those of the general land manager, and to that extent he must be thoroughly familiar with techniques of the agriculturist, range manager, and forester. His objective is concomitant with, not independent of, the primary use of the land.

Wildlife, like timber, beef, and corn, is a product of the land. It cannot be produced apart from the soil, waters, and vegetation which compose a landscape. There are some areas, as refuges, that are managed exclusively for wildlife; many other sites, as marshes, rocky outcrops, escarpments, gullies, and naturally infertile areas, are most productively used when made to yield a crop of wildlife. Because wildlife is a land crop, and its management to a large extent a cropping system, it takes its place along with the products of tilled fields, forests, and range in any scheme of land management. In attempts to use every parcel of land in accordance with its adaptabilities, it should be remembered that it is no more desirable to expend primary effort trying to raise wildlife on land suited to corn than it is profitable to attempt to grow corn on land best adapted to the production of wild vertebrates. It is an unwarranted waste of seed, fertilizer, machinery, and manpower to try to raise corn on land which will produce a poor corn yield, when that land could produce a profitable crop of wildlife.

The wildlife food patch serves to illustrate this idea. The plants used in food patches are usually ordinary crop plants, and thus they are almost invariably planted on good land capable of producing a tilled crop such as corn, soybeans, or wheat. From the standpoint of good land management, the farmer can realize most from his investment when such land is devoted to ordinary crop production, and it is a waste of land capital to use it otherwise. It is much more profitable to produce wildlife food on eroding streambanks, unproductive field borders, odd field corners, or other areas which it is economically unwise to use for more intensively managed crops. When established in a cover of adapted perennial plants, these areas, furthermore, become permanent parts of the farm pattern and landscape.

Wildlife managers are prone to err in their primary assumption, namely that since a given animal species can be produced on a particular piece of land, it should be encouraged there. It is much more fundamental to start by asking: What is the best crop this land will produce and should it be devoted to wildlife? Reference to our classification of land according to use capabilities once more serves as a useful guide. We learned in Chapter vi that Class viii land is useful primarily for wildlife. It embraces some extensive expanses of salt marsh and many smaller stretches of fresh-water marsh which will not become permanently productive cropland if drained. It also includes gullies, sand dunes, and acid bogs. These areas are usually large enough to be mapped on a scale useful for planning individual farms and ranches but, unlike other land classes, there are a great many parcels of wildlife land too small to be shown on a map of practical scale. It is the task of the wildlife manager to recognize such areas, whether mapped or not, and to plan their most appropriate management. That they are usually interspersed through areas of different cover type is all to the good.

Areas of wildlife land frequently unmapped are escarpments and rocky outcrops (Plate 26, p. 179), riverbanks and streambanks, irrigation and drainage ditchbanks, highly alkaline areas, and badly eroded spots. Such areas are widely scattered and many of them are small, although in the aggregate they loom large. Recent estimates indicate that wildlife lands on American farms and ranches add up to no less than 33,000,000 acres—an area as large as the entire State of New York (Bennett 1942; Davison 1942). Although no estimates are at hand, the total area best adapted to wildlife on non-agricultural land must equal or exceed that on farms and ranches. As previously stated, in so far as their profitable management is concerned, streams, ponds, and lakes should be looked

[172]

upon as Class viii 'land,' for they are best adapted to the production of a crop of wildlife, such as fish and waterfowl.

It is with this class of land that the wildlife manager has his prime responsibility, for it is his task to make these lands as productive as possible. It would be a waste of effort to try to grow tilled crops or pasture vegetation on them, and they are frequently too small, infertile, or otherwise unsuited to the productive maintenance of woodland. Yet, in any land-management scheme that attempts to use all the land wisely, these areas must not be neglected. The crop for which they are best adapted is usually some sort of wild plant or animal life. Some areas, like acid bogs, can be profitably utilized to grow special crops, like blueberries or cranberries. Others can readily be made to support upland game birds and small furbearers which can be benefited by the substitution of useful perennial plants for weedy or useless species, or by protection of existing desirable vegetation on such sites as: (a) unproductive strips between woodland and crop fields, (b) roadsides, (c) fence rows, (d) gullies and other eroding areas, (e) highly alkaline or otherwise infertile odd spots, (f) irrigation and drainage ditchbanks, (g) spoilbanks, (h) streambanks, (i) marshes and swamps which cannot be economically drained, and (j) rocky outcrops and escarpments.

Although 'wildlife land' is that on which the wildlife manager has primary technical responsibility, he cannot rightfully neglect other classes of land. For those lands which are devoted to the cultivation of tilled crops, i.e. Classes i, ii, iii, and iv, there are wildlife-management practices which can be established as integral parts of the land pattern. For Class i land, hedges are frequently applicable (Plate 27, p. 194); for Class ii, field border treatment; for Class iii, shrub buffers for wind erosion control as in the Great Plains; and for Class iv, pond-management measures. There are some types of Class i or Class ii lands so intensively used for crop production, as the blacklands of central Texas, where fences, field boundaries, and woody plants are absent, that the wildlife manager can contribute little or nothing to them, but they are rare.

Land Classes v, vi, and vii are dedicated either to pasture (or range) or to woodland. Wildlife-management practices applicable to grazing land or woodland are, therefore, applicable to these classes of land. On all of these, the biologist must work closely with other land-management technicians—farmer, agronomist, range manager, forester—and fit his recommendations into those for the primary use of the land.

There are many ways in which the wildlife technician can work with the land operator to obtain conditions suitable to wildlife. The farmer may be willing to defer pasture grazing until quail have hatched, or purposely preserve den trees for raccoon in his woodland. Measures that are of especial value to wildlife production, however, consist more importantly of so-called wildlife-management practices, which can usually be related to land classes or groups of land classes. Table 6 shows a few such practices and the land classes to which they are most applicable (Graham 1943). We need still more experience to be very specific about such a tabulation, but it serves to suggest that even though some wildlife practices are limited to Class viii land, more intensively used land classes can also support practices useful to wildlife. Such wildlife practices, of course, must contribute to or be compatible with the other land-management practices employed.

Thus a systematic consideration of land serves to give the wildlife manager, like other land-management technicians, a picture of the job he has to do. It serves as a foundation upon which he can build the quality and intensity of his work. The wildlife technician must weigh the results of his efforts on each class of land against the time, funds, and material devoted to making each class productive of wildlife. It sometimes happens that certain types of Class viii land, upon which the wildlife manager has prime responsibility, produce less wildlife, per dollar expended upon them, than Class i or Class ii land, upon which the biologist treats comparatively little acreage. The very high populations of ring-necked pheasants in northwestern Ohio, where grain and corn fields extend endlessly across a flat expanse, for example, produce high

[174]

WILDLIFE

TABLE 6. WILDLIFE MANAGEMENT PRACTICES AND LAND CLASSES

PRACTICE	LAND CLASS							
	Cultivated Land				Pasture or Woodland			Wildlife Land
	I	II	III	IV	V	VI	VII	VIII
Hedges....................	x	x	x	x	x	x	x	
Field borders..............	x	x	x	x				
Shrub buffers (wind erosion control)................		x	x	x				
Management of odd spots....			x	x	x	x	x	x
Pond management..........				x	x	x	x	x
Irrigation and drainage ditch-bank management........					x	x	x	x
Streambank management....					x	x	x	x
Spoilbank management......							x	x
Beaver management........							x	x
Marsh management........								x
Number of practices applicable to each land class....	2	3	4	5	5	5	7	7

wildlife returns on intensively cultivated (Class II) land where the wildlife manager has expended little effort. The pheasants nest and find shelter along the vegetated drainage ditchbanks from which they can forage into the fields for the spare grain which serves them as food.

[175]

Class VIII land itself varies in productiveness, for muskrats from a tidewater marsh provide greater economic return than shore birds on a coastal beach. The latter, of course, may have compensating values of a less tangible nature, as satisfaction to bird lovers. It is for the land-management biologist to determine the opportunities the land affords in the way of wildlife habitat, whether it is devoted primarily to the production of clean-tilled crops, pasture, or woodland, or is to be devoted entirely to whatever wildlife it can be expected economically to provide.

By appropriate attention within each land class to the habitat elements which the wildlife manager is in a position to influence, and by close working arrangements with land managers in other technical fields, almost every kind of land can be made to support beneficial wildlife adapted to it. All of this, like production of other crops, should not be undertaken just to produce 'more wildlife,' but should be commensurate with the costs involved and guided by the demand for the commodity produced, two items too little regarded by many who wish to encourage wildlife production.

In the past, many large tracts of land unsuited to cultivated crops have been settled and farmed in vain attempts to gain a livelihood from the tilled soil. It is now realized that submarginal areas are often more wisely managed when utilized for their range, forest, wildlife, or recreational values. The conversion of such land to uses other than crop production usually brings great advantages to wild birds and mammals, and the management of such land profitably includes attention to the wildlife resource. In accordance with concepts of sound land use, numerous submarginal areas scattered throughout the nation have now been readjusted to non-agricultural use and many of them are proving to be excellent homes for range, woodland, and marshland wildlife.

Just as large tracts once misused by man can profitably be converted to wiser and less intensive use that is beneficial to wildlife, so small, severely eroded sites may be made to serve wildlife also. The revegetation of gullies, galled spots, and other eroded areas with plants useful to wildlife has been so widespread that many persons think of gully planting as the chief erosion-control wildlife-management endeavor. Adapted trees, shrubs, and herbaceous plants have been variously used in clothing such areas with protective vegetation suitable for food and cover and such rehabilitated sites are known to be much used by many kinds of wild birds and mammals. On almost every farm there are small eroding areas standing within pastures or crop land that are too small to be effectively managed as woodland. There are rocky outcrops and small corners of fields which, for one reason or another, are not adapted to crop, pasture, or woodland use. Slight attention, perhaps only the prevention of burning and protection from grazing, transforms these areas into habitat elements that increase the interspersion of convenient cover and food.

ALTHOUGH the land manager is not primarily concerned with aquatic habitats, the treatment of some water areas is so closely related to the use of land that this chapter is devoted to a brief consideration of fresh water streams and ponds. A great deal of recent work has been done on the management of streams, lakes, and ponds for the production of sport and food fish, and useful reports are scattered throughout the literature for the fish manager. In so far as the land manager is concerned, protection of the soil resource and advantage to the land owner are of prime concern, and management of a stream or pond should fit into other approved types of land use.

Rivers and streams in the United States frequently have been looked upon as convenient places for the ready disposal of trash, sewage, and other waste materials rather than as natural resources productive of great material returns if given a degree of protection and some measure of thoughtful care. Throughout the United States, one of the most neglected parts of the landscape is the

---

PLATE 25

TOP. A typical transition area in South Carolina. Throughout the Southeast, the edges of fields adjoining woods produce a poor crop, erode badly, and provide a place for woody vegetation to invade the field. Useful permanent vegetation that will withstand the competition of the trees is needed.

BOTTOM. Three seasons later a stand of *Lespedeza bicolor*, a good honey plant and a preferred food of bobwhite quail, grows against the trees. Next to the grain is *Lespedeza sericea* which provides wildlife cover and a turn row for teams working the field. Such a perennial border exemplifies a wildlife management measure economically feasible, because it is a permanent part of the farm pattern and contributes first of all to the needs of the land.

stream margin or river bank. Ordinary land classification and mapping are not detailed enough to show that, although a stream may run through fertile Class I bottomland, its banks often belong in Class VIII, if they are eroded with infertile subsoil exposed, and that the banks frequently require treatment quite different from the land through which the stream flows. Crumbling streambanks threatening productive fields demand attention, and usually present a problem of community interest. An individual's efforts to stop freshet and flood destruction are wasted unless stream channels are controlled for long distances, and the entire watershed is protected by appropriate soil and water conservation measures. So long as the watershed feeding a stream lacks the protection of adequate vegetation, there will be erosion, silting, undercutting of streambanks, and drastic changes in habitat, not only in the stream itself but on adjacent land areas. One need not look far to see water courses cutting into fertile bottomland, and recovery of such raw, eroding areas becomes an important part of a reasonable program to protect a watershed.

So serious is this problem that some states are purchasing or leasing strips of land along the banks of major streams, to be publicly managed much as highway fills and cuts are managed to protect them from erosion. Protection by fencing and the planting of willow mats, willow cuttings, and other vegetation are useful bank control measures along streams of moderate size (Plate 28, p. 195). Even on large river banks, vegetation can be usefully

---

PLATE 26

TOP. Among those areas adapted to yields of wild plants and animals but not economically productive of woodland, pasture, or tilled crops, is the rough or odd spot illustrated by this steep area in an Ohio field. Protection from fire and grazing transform it into a haven for farm wildlife.

BOTTOM. In semi-arid areas water often determines the occurrence of wildlife. This Texas pond provides a fenced sanctuary for waterfowl and may support a few fish and furbearers. The pond consists of overflow from the tank in the foreground which provides water for livestock. The water is supplied by the windmill from the tower of which the picture was taken.

employed to supplement revetments, dikes, levees, and other engineering structures. When protected by permanent vegetation, the stabilized channel not only safeguards adjacent land by preventing erosion, but decreases siltation resulting from the crumbling banks of the stream. The stream is also improved as an aquatic habitat, and its vegetated margins provide an unexcelled environment for many kinds of wildlife. Fenced lanes to the stream for watering stock, of course, may be necessary where the stream flows through pastures.

As an erosion-control measure, beavers have been stocked in streams of the western states, especially in forested headwaters where they are not likely to damage crops or orchards and their impounded waters will not flood valley roads. The dams built by these mammals have been highly effective in conserving water and retaining silt. In the State of Washington, engineers estimated that one dam, constructed by beavers in less than two years' time, could only have been duplicated by human construction at a cost of $2500 (Scheffer 1938).

Beavers are soil and water conservationists of long experience and the occurrence of many natural meadows throughout the United States is now recognized to have been the result of beaver activity (Plate 29). In fact, the reintroduction of beavers to western regions, where they were originally abundant, has caused mountain meadows once more to produce a natural crop of hay as a result of the increased soil moisture retained by the dams the beavers built. (Plate 29, p. 210.)

STREAM POLLUTION

One of the most damaging influences upon streams as productive features of the landscape is pollution. Pollution from erosion silt, industrial operations, and sewage all do serious harm to streams.

SEDIMENTATION

The effect of man-induced erosion upon the environment is not limited to the land. Its influence upon aquatic habitats may be very great, for the waters of ponds, lakes, and streams are seriously modified by siltation, and sedimentation studies stress the rapidity with which reservoirs are being filled with silt and erosional debris. Although a great deal has been written about stream pollution by industrial wastes, not much inquiry has been directed toward the damage caused by suspended soil, which, however, is even more common and very probably more important as an influence upon aquatic organisms. Studies of erosion silt as a factor in aquatic environments indicate that it is a highly significant influence. M. M. Ellis (1937a) states that:

Erosion silt and other suspensoids (disregarding any specific toxic action of suspensoid wastes) affect fisheries directly by covering the bottom of the stream with a blanket of material which kills out the bottom fauna, greatly reduces the available food, and covers nests and spawning grounds; . . . the mechanical and abrasive action of the silt itself . . . may clog and otherwise injure the gills and respiratory structures of various aquatic forms, including many fishes and mollusks . . . Indirectly, but none the less effectively, erosion silt affects fisheries by screening out the light, by 'laking down' organic wastes, and thus increasing the oxygen demand at the bottom of the stream, and by retaining many forms of industrial effluents, as oils, chemical wastes, and pulps in beds on the floor of the stream, with disastrous results to the bottom fauna.

A factor so distinctly affecting aquatic habitats must determine the existence not only of specific organisms, but of aquatic plant and animal associations and their successional relationships as well. Two main successional trends of aquatic plant associations have been distinguished in English lakes, depending upon the presence or absence of silting by inorganic material. Studies of community relationships in aquatic environments as related to degree of sedimentation in the United States, however, have been much neglected. Nevertheless, the evidence at hand suggests that accelerated soil erosion may critically influence aquatic environments just as it does terrestrial habitats. It is reasonable to suppose that terraces, readjustments in land use, and other soil-conservation measures improve fish habitat, for even contour cultivation contributes toward keeping sediment out of streams, ponds, and lakes by reducing silt-laden runoff.

It is also well known that when erosion fills pools with soil or, by undercutting shady vegetation along the banks, permits the water temperature to rise, a stream once ideally adapted to trout can be converted into an environment totally unfit for such species of fish. A flourishing button industry once depended upon fresh water mussels taken in clear mid-western streams. It has now practically disappeared, reputedly due in large part to the fatal effect of erosion silt upon the mussels. The oyster beds of Chesapeake Bay and other coastal areas are also believed to be materially injured by excessive sedimentation. The effect of erosion silt upon fish populations has been emphasized by Langlois (1941), who contends that land-use practices of the past few decades have greatly added to the silt load of streams entering Lake Erie, increasing turbidity of the water and eliminating the dense aquatic meadows that once prevailed in the southwestern shore bays. Silt has been carried into the lake and deposited over the hard bottoms, completely transforming them as fish nesting sites. The result has been reduction of fish requiring vegetation for spawning and early growth, such as the yellow perch, and those needing clean, hard bottom for spawning, such as ciscoes and whitefishes. In their places are fish that will tolerate turbid water, such as sauger, sheepshead, catfish, and carp. That the change is detrimental to man needs no telling to those who have eaten Lake Erie whitefish or who have profited by the trade of the fishes originally inhabiting those unpolluted waters.

INDUSTRIAL POLLUTION

In spite of the damaging effect of erosional debris upon aquatic habitats, other types of pollution are sometimes considered to be even more serious. One authority (Ellis 1937b) contends that:

Although erosion silt is clearly the major silt problem in inland waters, locally various industrial silts constitute serious stream pollutants. Among such silts are included mine tailings, limestone, sawmill wastes, gravel washings, waste waters from ochre plants, and even blow-dust from cotton and wool cleaning establishments.

He might have mentioned also wastes from gas and dye works, paper mills, and chemical plants.

The idea that streams will clear themselves of any kind of pollution is false. They clear themselves only of certain easily oxidized wastes or those easily neutralized into harmless compounds, and then only if the stream offers sufficient stretch of open, agitated water. Furthermore, treatments of industrial waste that are satisfactory from the standpoint of human health, in that they eliminate injurious bacteria, may not remove the danger to fish, for the treatments frequently provide effluents which demand much oxygen, have a dangerous ammoniacal content, or even contain damaging quantities of salts. Chlorine, for example, will render water sterile by destroying bacteria harmful to man; it is fatal to most other aquatic organisms as well.

Modern methods of purifying streams must succeed in creating waters as free from materials injurious to fish and fish-food organisms as they are free of substances detrimental to human health. It should be remembered that pollutants may be injurious not only by poisoning water chemically, thus directly killing aquatic organisms both small and large, but also by covering stream bottoms, thus destroying nesting sites. Textile mill wastes suffocate fish by clogging their gills.

SEWAGE

How far we, as an enlightened nation, have yet to go in managing our ordinary affairs of life is shown by the National Resources Committee's *Report on Water Pollution* (1935), which states that although sewage systems are available to most of our urban population, only about half of all the persons in the United States are served by such systems. Of the sewage from our city areas, only about half receives both primary and secondary treatment. Thus many communities remain with inadequate sewage disposal or none at all, much sewage being discharged into our streams.

Last year a town of some 3000 inhabitants in the Texas panhandle inquired if it wouldn't be a good idea to stock minnows to keep down the bothersome mosquitoes that bred in a weedy, watery borrow pit along the highway just outside of town. The mosquitoes were living in the town's sewage, discharged into the borrow pit because the country was flat and there was no stream near by. Here proper urban planning and action, including installation of a proper sewage-disposal system, would have eliminated a troublesome land-management problem.

Stream pollution by sewage, however undesirable it may be from an aesthetic and health standpoint, may, if not too heavy, actually improve food conditions for fish by increasing the fertility of the water. Sewage contains a good deal of nitrogen, considerable phosphorus, and some other fertilizing elements. There are many instances where sewage, emptied into rivers and lakes, has increased fish yields.

Sewage more frequently, however, promotes conditions unsuited to fish life, especially when it results in a deposit of septic sludge. Its harmful effect upon fish through reduction of dissolved oxygen has been much considered. Carl L. Hubbs (1933) has discussed some of the other extremely deleterious effects of untreated sewage upon fish life. It may result in death of fry, ordinarily seldom seen so that this effect of the sewage is overlooked; it may also prevent spawn from hatching or cause development of malformed or stunted fish. Sludge may actually cover natural spawning beds, preventing eggs from hatching. Furthermore, sewage may kill the animal life on which fish feed or it may tend to drive fish away from polluted areas, even very small amounts of toxic substances agitating them and, consequently, making them easier victims to predators.

Some fish diseases are increased by sewage in the water, and so are parasites, such as the worms that cause black skin cysts, which live during part of their lives in snails, the latter being frequently more numerous in contaminated waters. Hubbs also points out an insidious relation between sewage pollution of streams and soil erosion. Heavy sedimentation accelerates the effects of sewage, for the fine soil particles pick up sewage particles by adhesion and settle to form a 'slowly decomposing sludgy mud of high organic content,' which suffocates the useful bottom life and renders the water less fit for fish life or for human use and enjoyment.

Ponds and Marshes

Many ponds are being constructed by farmers and ranchers throughout the country as a part of soil conservation work and agricultural land adjustment (Plate 30, p. 211). In ranch land, provision of well located stock watering facilities is a prime factor in promoting proper distribution of livestock on the range, and in the east there are few farms where a pond that supplies permanent water to pastured stock is not a management asset. The establishment of vegetation along the shores of a pond to prevent erosive wave action and its protection by fencing often prolong the useful life of the pond. Water for stock is frequently piped to a tank below the fence, or, on larger ponds in the West, lanes are fenced to a portion of the pond shore. The pond environs thus protected serve an additional usefulness as wildlife habitats (Plate 26). With slight further modifications, ponds can be stocked and managed for fish production. The pond may also supply water for irrigating the farm garden, help to maintain ground water supply, furnish ice, and provide an attractive place to picnic (Allan and Davis 1941). The restoration of eroded land frequently requires the damming of gullies, resulting in ponds which can contribute to the welfare and recreation of the farm or ranch family.

The value of ponds for migratory waterfowl—ducks, geese, herons, cranes—has been pointed out by Gabrielson (1939), who states that the 'maintenance of small bodies of water, the more numerous the better, is one of the positive actions looking toward the creation of recreational facilities and wildlife values that can be made a part of any flood control program, and with the minimum of cost.' Undoubtedly the establishment of numerous protected ponds and reservoirs throughout the central states, as part of a widespread land-management program, may do as much for waterfowl as larger, more widely scattered lakes. The smaller bodies of water present more 'edge' or margin for nesting and feeding areas, while the spread of waterfowl diseases is minimized where large concentrations of birds are avoided.

The experience of agriculturists shows that it is frequently uneconomical to drain marshes and swamps, for if it has not been determined that the soil beneath is productive, the cost of draining may be greater than the profit to be gained from the use of the drained land. If draining a marsh will not result in good cultivable land, such areas are best used when considered as wildlife land, for they can support waterfowl or afford a home for muskrats valuable for their pelts.

Much of the drainage work deemed essential to intensive agricultural operations in the East involves the construction of ditches to remove water from flat land productive only when properly drained. In the Southwest and on the Pacific Coast, drainage channels dispose of excess irrigation water, and in those regions water is supplied to farmland by means of irrigation canals and ditches. These drainage and irrigation ditchbanks are most satisfactorily maintained in some kind of vegetative cover that checks weeds, prevents the banks from eroding, and does not choke the channels. There are desirable perennial grasses, legumes, and shrubs that reduce ditch maintenance to a minimum, prevent erosion, and at the same time provide food and cover for wildlife. In intensively farmed areas, where corn, small grain, or some other crop seed usually is available adjacent to such vegetated ditches, agricultural land can support high numbers of game birds, as shown by heavy pheasant populations in northwestern Ohio. In semi-arid areas, irrigation and drainage ditches, when properly managed, can likewise serve as unusually favorable habitats for desirable wildlife.

FISH FROM THE FARM POND

About ten years ago at Auburn, Alabama, H. S. Swingle and E. V. Smith, a zoologist and a botanist, who were also fishermen, were trying a number of recommended schemes to improve fishing in a large pond, but without success. Finally they decided to try to discover for themselves what it was necessary to do to increase fishing in an ordinary pond.

On the grounds of the Alabama Experiment Station they arranged to set up several ponds in which to carry on their trials. They knew that the production of fish would probably depend upon the production of large quantities of phytoplankton—floating, microscopic plants. It seemed reasonable to try to increase the phytoplankton organisms by adding nutrient elements to the water. After a series of experiments in the laboratory, it was found that an ordinary commercial fertilizer, plus nitrate of soda, would result in a good growth of these minute plants. In that region, 100 pounds of 6-8-4 (N-P-K) fertilizer, plus 10 pounds of nitrate of soda per acre of water surface, is used to produce the amount of phytoplankton required. Fertilizer is applied as soon as danger of spring floods is past, and continued throughout the summer, at intervals of three or four weeks (Plate 29). The need for an application of fertilizer is determined by noting whether the bottom of the pond near shore can be seen through a foot and a half of water. If the bottom can be seen, then there is not enough plant food present. If, on the other hand, the water is murky-green or brownish with microscopic plants so that the bottom is invisible, then sufficient plant nutrients are present.

The microscopic plants serve as food for many minute animals and aquatic insects. These in turn form the food of certain kinds of fish called forage fish, as bluegill bream, which in their turn provide food for carnivorous fish, such as bass. Experimental work has shown that a balanced pond should contain a ratio of forage to carnivorous fish, for example, bream to bass, of about 3 to 1 by weight. This stocking, plus the application of fertilizer three or four times during the summer, is sufficient to produce 200 to 400 pounds or more of fish per acre of water surface per year. A pasture of average quality will produce annually 150 pounds, live-weight, of cattle or sheep, which dress out at approximately 55 per cent of liveweight. Fish dress out at 65 per cent. Thus an acre of pasture may produce 85 pounds of beef or mutton, an acre of fertilized pond more than 250 pounds of clean fish.

According to this method, which has been successful with ponds from a fraction of an acre up to 70 acres in size, the cost of production, exclusive of the original cost of the pond, is 3 to 10 cents per pound of fish. The cost of constructing a pond of about two surface acres is $100 to $200. On agricultural land, the production of fish for food from ponds is becoming increasingly popular, and farmers can well look to the farm pond for reliable sources of table fish in this country.

The pond must not be so shallow that it is likely to become choked with submerged or emergent water plants, and it should have a drain, so that the fish in the pond can be removed and the pond restocked if necessary. It is also essential, especially in the South, that the shores of the pond be kept as clean as possible, since the larvae of malaria-carrying mosquitoes develop in shallow water, particularly where there is debris or plant material in which they can hide. If the pond is kept clean and free of debris and higher aquatic plants, the bass and bream will have a chance to dispose of the mosquito larvae.

The method by which pond fish are produced may be summarized as follows. The fertilizer supplies the necessary plant nutrients for the development of large numbers of simple, minute algae. These are fed upon by insect larvae, especially those of midges (Chironomidae), as well as various types of protozoans, rotifers, and crustaceans. These in turn are fed upon by the bream, which, along with the insects, furnish food for the bass. Both the bass and bream are very palatable fish, providing food for man.

It is essential that the numbers of fish be kept from increasing too greatly, for otherwise there will not be enough food for them all to grow to good size. They must be harvested in a regular manner and as many fish as possible should be caught throughout the season. It is not likely that the pond can be over-fished by hook and line, for when there are too many fish for the food available they readily take the hook, and when the fish are reduced to the point where food is plentiful they no longer bite. After a season of fishing, the fish remaining in the pond are usually in the proper balance to produce a like quantity of fish the following year.

Here, then, is an example of a practical system of pond-fish management based upon a knowledge of the food chain involved, which may be diagrammed:

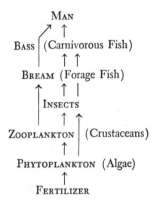

From a land-management standpoint, the significance of this type of fish production in the South is apparent when one recalls the fact that diet deficiencies are especially evident there. Increased quantities of animal proteins are of great benefit to rural populations in the southern states, and pond fish, such as bass, perch, and catfish, are higher in protein content than beef, pork, or mutton. Obviously, the value of ponds in the production of supplementary food for any farm table is considerable, and from the sporting standpoint the assured production of bass and bream is of interest to any fisherman.

Experimental work has also shown that the application of fertilizer to water in fall and winter induces a dense growth of filamentous algae, especially *Spirogyra* and long-filamented species of *Oedogonium*. These algae grow over submerged water plants, weighing them down and screening them from the light, usually killing them by spring when the algae, which are winter annuals, have disappeared. Here, then, is a biological method of controlling objectionable weeds that is not detrimental to fish-pond management. The scheme has been used to rid Alabama ponds of various undesirable plants in the genera *Potamogeton*, *Najas*, *Chara*, *Nitella*, and *Myriophyllum*. Larger emergent plants like water lilies and spatterdock are best removed by dragging from the pond or cutting their leaves repeatedly (Swingle and Smith 1942). Deepening the edges when the pond is constructed also tends to reduce the marginal shallow area in which the larger plants usually grow (Davison and Johnson 1943).

That fish can be produced as a part of an ordinary crop rotation is a long-tried practice scarcely known among Americans. Throughout the upper Rhone Valley of France, farmers grow wheat or barley in a field for a year or two, then, if it is not in a natural depression, dike up the edges of the field and fill it with water. The decomposition of the grain stubble, together with any fertilizer that may have been applied to the soil, produces sufficient micro-organisms to sustain a food chain with stocked fish at its head. Waterfowl are also supported by these ponds.

After a couple of years under water, the field is drained, and the remaining aquatic vegetation and unutilized fish serve as fertilizer for the small grain once more planted in the field. Such fields vary from a few acres up to several hundred acres in size, and generally do not hold water to a depth of more than 4 feet. The fish are usually carp. Thus there has developed a rotation of land farming with water farming productive of both bread and meat, characteristic of efficient management of the land by the French farmer.

LAND MANAGEMENT AND THE WATER SUPPLY

One of the less easily recognized effects of certain types of land mismanagement is the effect upon sources of domestic water supply. It is well known to many who have lived close to the land that, as cultivation increased, forests were felled, and grazing was intensified, many springs, free-flowing when the land was new, dried up and became useless. Beaver meadows, once lush with native grasses, no longer produced hay when the beavers were gone, and with draining of swampy areas near by, ponds and lakes have completely disappeared. The draining and burning of the Everglades has threatened the water supply of east-coast Florida cities, with no profitable results to the land drained; other examples will occur to the reader.

What has happened in the Santa Clara Valley, extending southward from San Francisco Bay in California, illustrates the way in which man's use of land can affect the ground water supply (Bird 1942). Surrounded by mountains, the valley occupies 200,000 acres which were at first grazed and dry-farmed by the Spaniards, who settled there in 1777. It was not long after the Americans arrived in the 1850's, however, that artesian water was discovered and put to use irrigating fruit and the other crops which thrived there. By 1910 there were 1000 flowing wells. In the meantime, pumped wells were sunk into the valley fill above the artesian belt, the average pumping lift originally being only 35 feet.

Nothing, however, was done to regulate the flow of the ground water, nor to replace it as it was used. Much of the water that originally percolated into the soil from the intermittent storm-fed streams rising in the surrounding mountains was permitted to run off in damaging floods. Measurements in 1931-2 showed that only 29 per cent of the runoff from the watershed went into underground storage, all the remainder went to waste in San Francisco Bay.

Pumping expanded so rapidly that by 1933 the water drawn from the ground was more than 5 times the amount used 20 years previously, with the water table dropping about 5 feet per year. In 1933 alone the water table dropped 21 feet and by that time it had sunk an average of 130 feet over the entire valley. The pumping lift was 165 feet and irrigation became a precarious and costly operation. With the decline in water level, artesian wells gradually played out; the last one closed in 1930. Two thousand pumps were then pulling water from the valley.

With the removal of water from the valley fill, the ground slowly settled—sinking five feet in 20 years. Damage was caused to buildings, pipe lines, and streets at a cost of millions of dollars. Worst of all, the storage capacity of the underlying geological formations that carried and yielded the supply of water was permanently reduced.

At least part of this loss of a valuable resource might have been spared. Technicians familiar with the situation warned the public of the disastrous result of their continued heavy use of the underground water supply, which many thought to be inexhaustible. In 1922, a $4,000,000 conservation plan was defeated by a 7 to 1 vote, yet in the following 20 years more than $16,000,000 was spent for new wells, more powerful pumping equipment, and increased power. By 1934, when water was no longer available in parts of the valley, and some wells began to pump salt water from the Bay, a program was finally started to hold and control flood waters from the mountains, build percolation reservoirs, canals, and water-spreading beds, and otherwise replenish the ground water supply. Thus the water table has been raised some 65 feet so that pumps need to lift water only 85 feet—an improvement, but a far cry from original conditions or what might have existed if careful management had been undertaken as soon as the fault became apparent.

There are very few parts of the United States, even the more humid sections, where rainfall during the growing season is sufficient to support the crop that is grown. Growing crops in large part depend upon water stored in the soil, subsoil, and parent material beneath. From this source in the soil, wells fill with water

for human use, as do springs and running streams. The maintenance of a large ground water supply is essential to human welfare. It is almost axiomatic, therefore, that any type of land management which prevents too rapid runoff but permits maximum
percolation and infiltration of precipitation is highly desirable.

There have always been floods, even in areas once covered by
forests, and probably there always will be floods. It is known, however, that increase in volume and rapidity of runoff gives abnormally high and steep-fronted flood waves, increased rates of peak
discharge, and greater velocity to the flood waters. These factors
are responsible for the greater transporting and eroding power of
water as well as for bank overflow, and anything which tends
to reduce them not only aids in preventing disastrously damaging
floods, but helps to maintain a useful supply of ground water.

Important as great dams, levees, and other structures can be in
helping to prevent loss of water, it is to the management of land
that we must look first for conservation of our fundamental fresh-
water supply. When the lands best suited to forests are clothed
with a well-developed stand of trees, when the western range and
eastern pastures are in a good cover of grass well balanced with
the livestock that use them, and when crop lands are managed
in accordance with most modern soil and moisture conservation
practices, we shall have gone a long way toward preserving a
natural water supply without which we cannot get along.

EXOTIC THINGS

XII

THE PROBLEM

A MATTER in which the practical ecologist is much interested is what happens when non-native plants or animals are introduced into a region. In older countries, such as England, some sort of biological balance has been attained under man's long use of the land, but in newer countries the consequences of introductions are not easily predicted and may present extremely complicated biological interactions. The frequently told story of the zealous gardener who wanted to make Hawaii even more beautiful than it was is worth repeating in this connection. He introduced there an ornamental tropical American shrub, *Lantana camara*. In its home, this plant 'knows its place' but, as we shall see, in Hawaii it took full advantage of its new environment.

Sometime before the introduction of this shrub, turtle doves from China had been brought to Hawaii, and Indian mynah birds also were introduced. Unlike natives, these two birds fed heavily

---

PLATE 27

TOP. This South Carolina hedge is composed of low-growing plants which do not shade or interfere with adjacent crops. Consisting of selected fruit-bearing species which afford foods for home use, they also provide food and cover for wildlife and serve as a permanent guide for marking contour cultivation.

BOTTOM. Hedges serve to interrupt the flow of water across long slopes, retarding its force and checking soil washing. Note the silt banked against the shrubs in this poorly managed corn field. The hedge, however, serves best as a supplement to other erosion-control practices, like the contour tillage shown in the top photograph.

upon lantana fruits. The aggressiveness which the plant displayed in its new habitat plus the capacity of the exotic birds to distribute the seeds combined to make the plant a serious pest in parts of the islands devoted to grazing. But there is even more to the story. Before the mynahs were introduced, the Hawaiian grasslands and young sugarcane plantations had been severely damaged by armyworm caterpillars. When the mynahs came, however, they helped to keep the armyworms under control.

Meanwhile someone got the idea that certain foreign insects would check the spread of lantana by eating the seeds. Consequently, insects were introduced. As predicted, they destroyed so much seed that the lantana began to decrease. Then the mynahs, deprived of lantana seeds for food, likewise began to decrease. This resulted in a recurrence of armyworm outbreaks. Furthermore, many of the places now vacated by the shrubby lantana became occupied by other introduced shrubs, even more difficult to eradicate. Here is a lesson about the reckless introduction of exotics, for the result in this instance was an ecologically unbalanced situation. Only by a long, difficult procedure can some sort of desirable stabilization be brought about.

In the United States much damage has been caused by exotic insects, such as the Hessian fly and cotton boll weevil. A great many of our most pestiferous plants are also foreign. Of course, native species may cause trouble, as do the grasshoppers of the Great Plains. In defense of introductions, one might argue that many of the most important cultivated crops of the United States

---

PLATE 28

The eroding creek or river bank which cuts into fertile bottomlands presents a serious land-management problem. Along this stream in west-central Wisconsin (top) the tumbling banks were sloped, willow poles were laid in upright trenches, and willow bulkheads were constructed (center). Three years later the willows were so well developed that they obscured the view from the same point on the bridge from which the pictures above were taken (bottom).

and nearly all of our domesticated animals are non-native species. For the most part they survive, however, only when tended carefully by man. This discussion is not intended as a statement to the effect that introductions are necessarily harmful, but rather to emphasize the fact that great care must be exercised in tampering with the natural condition of an area, and that it is imperative that we think ecologically as well as in terms of simple cause-and-effect relationships.

When a species is introduced into an area where it has not lived before, it is almost impossible to foretell the consequences, although it is quite probable that it will either succeed gloriously or eventually fail entirely. In a study of the birds introduced into the United States, Phillips (1928) substantiates this fact. Some birds, like the ring-necked pheasant when stocked in the Corn Belt states and the European partridge in the Northwest, increase rapidly, spread, and become very much at home. It is interesting to note, however, as mentioned below, that these same birds have never survived in some parts of the United States, even after repeated attempts to establish them. Other introduced birds never bred and quickly vanished. Such were the capercailzie, black grouse, and many European songbirds, like the nightingale. Attempts to establish our own California quail and pinnated grouse in the eastern United States have also failed. Some species, as the European partridge, when stocked in the Atlantic Coast states, bred but gradually disappeared, and still others, like the European skylark and European goldfinch, bred and gained local footholds in the eastern United States for some 20 years, but then gradually, or after a severe season suddenly, disappeared. Some birds even become pests before they disappear, as did the California quail and ring-necked pheasant in New Zealand.

Among plants, what has been done with the Douglas fir of western North America provides an interesting lesson in transplanting. Douglas fir is one of the most valued timber trees of England and Europe, far superior to species native there. It grows fast, is adapted to a variety of soils, and even reproduces itself in the Old

World. Yet it is not every Douglas fir that does so well there, but only seeds of firs from the Sierras, the Cascades, and the Coast Ranges of westernmost North America. Seeds from Douglas firs growing in the Rocky Mountains, as in Colorado, persistently fail in Europe and England. Throughout the British Isles, Belgium, Germany, Switzerland, and other continental countries from Sweden to Portugal and Austria, it is the coastal form of the Douglas fir that makes an outstanding contribution to the forestry of those nations, while the Colorado form of the tree everywhere grows slowly, is generally inferior, and frequently fails.

In the eastern United States, on the other hand, if you wish to grow Douglas fir, you must be sure to obtain seeds from the Colorado forms, because in the East the coastal firs do very poorly and frequently fail. In the northern Rockies, where the forests of the Cordillera and those of the coast meet, the Douglas firs are intermediate in characteristics; seed from northern Idaho, for example, may as often fail as succeed in either the eastern United States or Europe. With inherent characteristics of living things so striking and persistent, those who wish to manipulate their transfer from one natural area to another will do well to proceed cautiously on the basis of well-advised trial and experimentation as well as to make use of all the tools of management the applied biologist can provide—which, unfortunately, are as yet very few with respect to this problem.

Although some plants and animals quickly succumb when taken to a new home, great difficulties have arisen from the transplantation of others. In the United States, serious plant diseases, like the chestnut blight, and many of our noxious weeds are exotics, among which may be numbered bindweed, yellow-rocket, Russian thistle, cheat grass, and water caltrop. In other countries it is as bad, as witness the muskrat in Europe, the European rabbit and water hyacinth in Australia, the prickly pear in Australia and the Mediterranean, and *Elodea* in England, all of which have created troublesome problems.

Good is seldom said of the benefits of introduced weeds, although use may sometimes be made of them as indicated by the following example from Australia (Zierer 1941):

In the upper Brisbane River and its tributaries navigation by pleasure boats and ferries is commonly handicapped by dense growths of water hyacinth. Its accumulation against bridge piers during floods frequently endangers such structures. Along with prickly pear, lantana, and cape weed, water hyacinth ranks as one of the important plant pests of Australia. During times of drought, however, water hyacinth is dragged from streams by dairymen as emergency feed for cattle. When the streams are in flood the hyacinth is swept into the tidal portions of the river and into the bay where it is killed by contact with salt water.

The wildfire success of many introduced species is not easily explained. Perhaps it is owing to lack of the species' usual enemies, because of some inherent characteristic of the species, or because it fits a niche where competition is at a minimum. The way in which weeds populate a garden only a few short weeks after one stops hoeing it is very little short of marvelous. It is scarcely less wonderful that many of these same weeds, if their seeds find their way hundreds or thousands of miles from the garden, will grow and rapidly reproduce their kind to populate almost any recently disturbed area.

Transplanted from both the Americas to Australia during the eighteenth century for the purpose of providing potted ornamentals and hedges around homesteads, prickly-pear cactus plants (Opuntia inermis, O. stricta, and others) became pests of unprecedented proportions on the island continent, occupying millions of acres of crop and range land. After many attempted methods of repression had failed, resort was made to biological control. Following tests to be sure that they would not attack plants other than cacti, a moth, Cactoblastis cactorum, and some other insects were introduced from Argentina.

Alan P. Dodd (1940) describes this example of biological control in a recent bulletin of the Australian Government. He shows that the conquest of prickly-pear in Australia has been accomplished

[198]

almost entirely by the *Cactoblastis* moth. Dodd states that its introduction brought a complete change within a few years; its progress was spectacular; its achievements bordered on the miraculous. Great tracts of country, utterly useless on account of dense growth of the cactus, have been brought into production as a result of the insect's activities, and the prickly-pear territory has been transformed as though by magic from a wilderness to a scene of prosperous endeavor.

One consignment only of the moth was introduced into Australia, 2750 eggs being sent from the Argentine in March 1925. After being reared in cages through two generations, the small original number had multiplied to 2,540,000 by March 1926. The first trial liberations were made in February and March 1926. Large-scale rearing in cages was continued until the end of 1927, by which time 9,000,000 eggs of the insect had been released at many selected places in Queensland and New South Wales. By that time, the increase at the earliest liberation sites had been so great as to render further rearing at the field stations unnecessary. A campaign of mass distribution was begun in 1928 and was completed in 1930, eggs being collected from the field centers where the insect had multiplied to very large numbers. The huge total of approximately 3,000,000,000 eggs was released throughout the prickly-pear territory.

The efficiency of *Cactoblastis* was apparent almost from the outset. The larvae were gregarious, tunneling in the segments and stems of prickly-pear, and reducing the plants to a rotting mass. Within 15 months after the first trial liberations had been made in early 1926, many large plants of prickly-pear had been destroyed at various points. By June 1928, where several of the early liberations were made, the insect occurred in great quantity; it had also dispersed over a considerable area, and had caused the destruction of dense growths of pear on hundreds of acres. The big distribution program of 1928 to 1930 resulted in general establishment of the moth. In 1930 to 1932, an enormous *Cactoblastis* population occurred throughout the prickly-pear territory, and the great areas

[199]

of the pest throughout Queensland and northwest New South Wales were reduced quite suddenly to decayed pulp. The rapidity of the onslaught was astonishing; mile after mile of dense growth collapsed in a few months under the concentrated attack of phenomenal numbers of the larvae. The last big area of original pear in Queensland succumbed to the insect in 1933, seven years after the initial liberations had been made.

Here, then, is a story of how an introduced species, carefully checked before it was released, served to control another introduction that had found its way into a new environment without adequate check on its potential behavior. It is probably the outstanding example of its kind. It has given rise among certain factions in the western United States to cries for introduction of *Cactoblastis* to decrease prickly-pear cactus where it is considered a range weed. Entomologists agree, however, that its introduction here would be a much different thing from that in Australia, where no cacti are native and the introduced ones were everywhere pests. In the United States, about 200 species of cacti occur, many of them of considerable ornamental value and some, like the giant cactus and organ pipe cactus, are protected by law as scenic features of national importance.

Furthermore, it would probably be impossible to prevent the spread of the insect into Mexico where the prickly-pear has been cultivated for centuries for its edible fruit, the tuna, of which the Mexicans have developed as many varieties as we have kinds of improved apples. In this country, the cactus never grows as lush, heavy, and thick as in Australia, and in places, as in southern Texas, some ranchmen consider it good emergency livestock feed. Thus an advisable biological control in one situation may be very questionable in others, and the land-management biologist must weigh each case in the light of the circumstances surrounding it.

The use of introduced insects, however, has been largely confined to the stocking of species predacious or parasitic upon other insects. One of the first attempts in the United States at biological control of a pest was the introduction more than 50 years ago of the Australian lady beetle (*Rodolia*) to control the cottony cushion scale, a serious pest of oranges in California. The principle proved so effective that it has been used subsequently in the control of other insect pests. If, however, a native insect pest is already subject to attack by numerous predators or parasites, such a means of controlling it may not be successful. Introduced insects have usually left their enemies behind; their natural enemies, if introduced, may or may not increase and attack them.

Biological control is relatively inexpensive and has been used successfully in the United States against not only the cottony cushion scale, but the black scale, California's most destructive pest of citrus, several mealy bugs attacking citrus, the woolly apple aphid, the gypsy moth, and others. Natural enemies have not been successful in the control of most aphids or of grasshoppers, one of our most devastating insect groups.

Biological control is not confined to insects. The mongoose, a small carnivorous mammal from India, was introduced into the Caribbean islands to reduce rats and mice, which it did so successfully that it had to turn for food to other small animals, both wild and domestic, whereupon it became a great nuisance. A comparable example among plants is the use of the perennial Asiatic crested wheatgrass, which is now planted throughout the northern Great Plains as a range improvement measure. The crested wheat not only grows well, furnishing good grazing, but has a tendency, at least for a while, to crowd out the range weeds, many of which are themselves natives of foreign soils.

It is interesting to note that not all introductions are likely to become pests. The Monterey pine, whose natural range is limited to a few scattered colonies along the southern California coast and adjacent islands, grows exceedingly well in Australia, where it produces saw timber in 25 years. We have also seen that the Douglas fir in Europe is very successful. Usually less can be said of introduced timber trees, however. Boyce (1941) has pointed out that exotic trees frequently have poor form, as shown by Scotch pine in this country. Scotch pine plantations in Pennsylvania and New

York grew well for 20 years, then were quickly decimated by fungi, spittle bugs, and Woodgate gall rust. Erosion-control plantings of black locust trees in Minnesota and Wisconsin, far to the north of the native range of the species, grew very well until a severe early frost destroyed them.

Introduced trees are often susceptible to diseases to which native trees are resistant, as proved by many attempts to establish North American trees in European forests. Boyce concludes:

Exotics are not all foredoomed to failure, but for every exotic the chance of failure is much greater than the chance of success, so that there should not be disappointment at the immediate or ultimate failure of an introduced tree but merely pleasant surprise if, after a century more or less of trial, an introduction is found to be a success. Although extending the range of a native species has more chance to succeed, yet the same considerations apply as to exotics. Finally, in the United States with its great number and variety of valuable species, there is less need for introduced trees for forest purposes than anywhere else in the world.

William R. Van Dersal, in his attractive and authoritative book *Ornamental American Shrubs,* has recently made a good case for the native shrubs of the United States, pointing out their particular adaptability for use as first-rank ornamentals in this country which is their home.

We are always reminded by those who want to introduce exotics, of course, that our very agriculture rests upon foreign plants. What would we do without wheat, barley, rye, buckwheat, pasture grasses and legumes that are not native, and our livestock, all of which, except the turkey, are foreign? To that, of course, we can reply that corn, American cotton, peanut, tobacco, sweet potato, white potato, squash, and many beans are American, and contribute as much to our economy as the introduced species. Anyway, the danger from introductions is not as threatening among cultivated plants and domesticated animals, which exist through watchful tending and artificially sustained environments, as it is with wild animals, forest trees, and range grasses.

On western grazing lands, crested wheatgrass, a Siberian species, is growing so well as a renovator of overgrazed areas, especially in the northern Great Plains, that it is being hailed by some people as the Messiah of the range. So long as it is profitable to reseed it or it can maintain itself against competition by native grasses, it will be useful. Otherwise we are probably wiser to look upon it as a temporarily valuable, but artificial member of the community, eventually replaceable by western wheatgrass and other native species as the range, once properly stocked, moves up into a higher successional stage.

With respect to the whole question of introducing a species, the ecologist will want to know: First, will it survive; second, will it spread widely, take the place of a now useful species, or otherwise seriously upset present conditions; third, is it more valuable than a native species now available which can become useful with appropriate attention; and, fourth, can it maintain itself successfully enough to compete with other species for its place in the community of which it is to become a part, or will it require the special kind of care given to crops and livestock? These things must be kept in mind, for as much harm as good can easily result from exotic species unmindfully introduced.

CLIMOGRAPHS AND MAPS

While examples of past experience are useful, they cannot show whether an untried plant or animal will succeed in a new environment. It would be well to have such a guide, for there are many instances where it is to our advantage to foster new things if we can safeguard the procedure. In the proposed home of the plant or animal to be transplanted, vegetation, climate, and other habitat factors should be generally similar to those of the original home of the species. This should be prerequisite, although such habitat conditions do not indicate effects of competition, or whether the new species is likely to force out of the particular ecological niche it will occupy, a valuable native species now in that niche. Will the Chukar partridge, for instance, which game enthusiasts are attempting to introduce in various parts of the United States, be able to compete with, or be likely to supplant, the bobwhite or the California quail?

An attempt to evaluate habitats with respect to transplanting living things has been made with a diagram called the climograph. Such a diagram is constructed by plotting the mean monthly temperature for a designated weather station against the precipitation for the same month. This is done for the twelve months of the year, the twelve points then being connected by a closed line. Strictly speaking, when such a diagram charts precipitation and temperature, it is termed a hythergraph; one charting humidity and temperature is a climograph. Either may be called a climate diagram, and the term climograph is now generally used for either type, especially since charts giving humidity are seldom used.

Such a diagram thus graphically represents, so far as rainfall and temperature averages suggest it, the habitat of a given plant or animal. If the climograph of the region from which a species is to be transplanted coincides, at least for critical periods, with that for the region into which the species is to be introduced, it is assumed, in so far as climatic factors are concerned, to be safe to make the introduction. The ecologist recognizes strict limitations to devices of this sort. The climograph, for instance, is not only limited in the factors it evaluates; it is subject to the same criticisms that are levied against all climatic data that portray broad averages only. Furthermore, a game animal may not be introduced from its native home, but from a region where the animal has already undergone a period of acclimatization and change.

Climographs were first used to determine suitability of Egyptian oases and Australian areas for human habitation. They have also been used to indicate world regions suitable for sheep production, success of cattle raising and chicken breeding, and, lately, for determining areas adapted to flax production. Recently they have been used as an index in the determination of natural vegetation types (Smith 1940). Animal ecologists have made considerable use of climographs and game managers have employed them to suggest desirability of stocking certain areas with exotic birds.

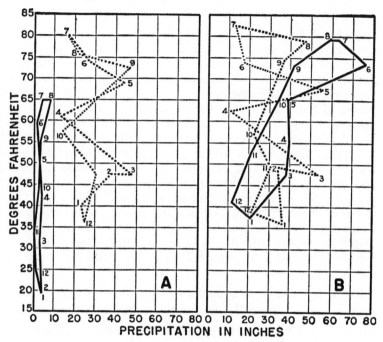

FIG. 8. Climate diagrams. Average monthly temperatures are plotted against total monthly precipitation, the months numbered consecutively from 1 to 12. In A the solid line represents conditions at Leh, India, the home of the Chukar partridge; the dotted line Knoxville, Tenn. In B the solid line represents Zi-ka-wei, China, home of the ring-necked pheasant; the dotted line Nashville, Tenn. (after Cahn, 1938).

One of the outstanding transplantations of game birds into the United States has been the introduction of the ring-necked pheasant. It has increased to astonishing numbers from northern Ohio to the Dakotas and the irrigated sections of Kansas and southern New Mexico. Attempts to introduce the birds in the eastern humid region south of a line running roughly from central Iowa to southern Pennsylvania, however, have consistently failed. The reason is not obvious, but Cahn (1938) offers, among others, a climograph to show that it is unreasonable to expect ring-necked pheasants to survive in Tennessee, where repeated attempts to introduce them have been unsuccessful (Fig. 8). Cahn states:

The climograph for Nashville . . . sprawls all over the climograph for Zi-Ka-Wei, China, which is in the heart of the native Chinese ringneck pheasant country. Yet note that in spite of this fact, only three months during the entire year lie within the native optimum range of the species: November, December and February. All of the months associated with incubation and the rearing of the young fall far outside of the tolerance range. This particular graph illustrates another point which should be mentioned in passing: the value of *mean monthly* data as compared with *mean yearly* data. If the mean annual temperature for Nashville and Zi-Ka-Wei were examined, it would be found that the average temperature for the year in Nashville is 54.8° and that of Zi-Ka-Wei 58.5°, a difference for the entire year of but 3.7°. On this basis it might be assumed that the climate in this region of China corresponds very closely with that of Nashville, yet the month by month analysis shows, as indicated, that only three months during the entire year fall within the actual conditions in China.

Cahn also charts climographs for the home of the Chukar partridge and of Knoxville, Tennessee, to show that the partridge will not survive in the Tennessee Valley because 'not a single month in the Knoxville yearly cycle remotely approaches the conditions represented in Leh, India.' Comparable studies of other birds have been made by Twomey (1936).

In attempts to introduce or stock non-native plants or animals, the land manager will do well not only to use such devices as the climograph, but to consult those maps of natural conditions that are available. Physiographic maps may aid in evaluating elevation, physical barriers, and uniformity of topographic features. More useful, perhaps, are vegetation maps which portray types of plant cover. Such maps obviously caution against transporting a species typical of deciduous forest, for example, into country that is predominantly grassland. Hardiness maps indicate conditions under which given plants are likely to grow well, and plant-growth region maps are practical guides for predicting the areas within which a plant species may reasonably be expected to survive. As suggested in a preceding chapter, plant indicators can also be used, and throughout the world many plant species have been transplanted successfully from one place to another because some plants with which they had been associated previously were known to thrive in the new locality.

As noted in the chapter on wildlife, however, it is not upon the introduction of new and unusual animals, or plants, however attractive and novel they may be, or, in the case of game species, upon the artificial stocking of pen-raised birds that we should depend for supporting a population. Greater attention to environmental factors will assist in the conservation and production not only of wild species, but of domestic ones as well.

What Ralph T. King has recently written about the wisdom of introducing exotic game birds is a useful guide that might well be applied to other kinds of animals and plants. In summary, King (1942) draws the following tentative conclusions:

1. The introduction of foreign species has been highly expensive.
2. The number of established introductions that have proved to be non-beneficial or actually injurious is as great as the number that has proved desirable.
3. Introduced species, once established, create demands on both food and cover, may spread new parasites and diseases, and may result in cross-breeding to the detriment of closely related native stock. Introductions do not necessarily result in reducing the hunting pressure on diminished native species.
4. Introductions into exhausted or deficient environments can only result in loss of the animals and greater deterioration of habitat elements.
5. Introduced species may increase to pest proportions.
6. Population behavior, food habits, effect upon native vegetation and fauna, and spread of an introduced species is uncertain until several years after the species is successfully established.
7. Unfortunately we have learned relatively little about the costs and results of introducing species into this country, thus failing to profit by past experience.

CONTROL

XIII

THE task of suppressing undesirable forms of life presents a challenge to our knowledge of interrelationships. Many of us are most familiar with control methods, perhaps, as they relate to human diseases. In the light of some of the ecological or 'naturalistic' methods of control discussed below, a word about diseases is warranted. Even with disease-producing organisms harmful to man, we do not always start out to destroy the organisms believed harmful, and many human diseases are controlled only when all the relationships involved are understood. We frequently prevent a disease, for example, not by directly fighting it off, but by orienting ourselves with respect to our environment, as keeping out of draughts and avoiding exposure to prevent colds, or shunning those whom we know to be infected with a contagious disorder.

We prevent some diseases by controlling the host of the disease-carrying organism, rather than by directly attacking the organism itself. The story of malaria is probably worth briefing as a useful parallel to the way in which the land manager must approach control problems.

When the Romans decided that malaria was caused by the night air, they ordered all the doors and windows of their homes tightly closed from sundown to sunrise to exclude the 'pestilence that walketh by night.' Little did they realize that more than 1500 years would pass before the truth would be known about malaria —that the disease is carried by the *Anopheles* mosquito. Yet of all the plagues of human history, malaria probably has taken one of the heaviest tolls of human lives. For 2500 years it has been recorded throughout many parts of the world. Before the rise of Rome it was recognized as the most deadly enemy of the Athenian Empire, and if she had conquered malaria, Athens might well have ruled the world. Ancient Rome lost more soldiers to malaria than to her enemies.

The Roman authorities tried diligently to discover the cause of this plague, and came near the solution, for they found a relation between the disease and the night air, and closing their homes at night was correct procedure. But the next step in man's knowledge of this malady was delayed 15 centuries until someone began to look for further relationships. Even then, long, patient experiment was necessary to learn the baffling facts that infective transmission requires a certain interval of time, that only a few of more than 100 species of mosquitoes can transmit malaria from one person to another, and that of these it is the female alone which carries the disease.

Until the vector was found, attempts to control the organism actually causing the disease could only be random ones. Such an example is worth learning well for it teaches a lesson applicable to the control of many kinds of troublesome and injurious animals and plants, large and small—that most effective results are not always obtained by the most obvious course.

Frequently attempts to control injurious rodents, harmful insects, and weedy plants have been positive methods representing direct attack. Such methods are not necessarily permanent or, in the long run, the most economical. Damage by noxious plants and animals is usually a result of their overabundance, and it should be remembered that there are ways to control high populations of harmful species in addition to such aggressive methods as poisoning, trapping, shooting, spraying, or gassing.

Nonaggressive control of injurious species, like means of increasing useful forms, is likely to be most successful when it is related to the use of land. For example, suitable tillage and regulated grazing, as well as controlled clearing, burning, or flooding, are cultural methods useful in reducing injurious species, while biological controls, such as change of habitat, use of living barriers or buffers, and the encouragement of natural enemies and diseases frequently prove successful. Other methods, more direct and usually more costly than cultural and biological controls, are mechanical ones such as use of non-living barriers (e.g. fences), proper design of dams, terraces, and other engineering structures, and chemical practices such as the use of repellents and sprays.

When the occurrence of an unusually large number of individuals of a species is caused by a natural cyclic fluctuation in the population of that species, its abundance is rarely as troublesome as a high population which results from a change of environment brought about by man's influence. Pronounced cyclic fluctuation is not known to occur in a great many animals, but when it does occur in a species of concern to the land manager, it is usually most economical to correlate operations with phases of the cycle.

From the 1936-7 *Annual Report of the Bureau of Animal Population* at the University of Oxford, England, a statement on control is worth quoting:

There is no human being who is not directly or indirectly influenced by animal populations, although intricate chains of connection often obscure the fact. Population problems are as much part of the fabric of daily existence as is the weather, and equally important. What is different is that not only do animals have this influence on man, but man has an increasing power over the

---

PLATE 29

TOP. Scattering commercial fertilizer from the edge of a farm fishpond as a 'land'-management measure in the production of pond fish. New techniques make the farm pond as productive of food, acre for acre, as well-handled pasture, and emphasize attention to water areas as a part of a comprehensive land-management scheme.

BOTTOM. The mountain meadow beside this beaver dam in Oregon owes its existence as much to biological agents—the beavers—as to physiographic ones. In many parts of the West, when the beavers were removed, lush meadows disappeared and erosion followed the flash runoff no longer impeded by the check dams the animals had once maintained.

fate of the animal populations that still throng the world. There is less of a moral problem about going out on a doubtful day without an umbrella, than there is in ordering the destruction of a species on the chance that it may be doing harm to human interests.

This from an agency that has had a great deal of experience with the subject and which we in the United States might do very well to emulate. Americans have not assumed it important enough to spend much effort learning of animal populations. We have few universities and no governmental bureau which undertake specific research on numbers of vertebrates as they relate to land management and human welfare. It is of more than passing interest that the Union of Soviet Socialist Republics considers the study of population dynamics so important that well-planned and amazingly extensive ecological studies of rodents are being undertaken throughout Russia. With respect to diversity of rodent fauna and types of land-management problems, their country is not greatly different from ours.

Methods of Control

Methods designed to minimize damage from living organisms destructive to vegetation or harmful to useful animals should be based upon sound biological principles, whether the end to be achieved is the control of a rodent or the checking of a plant disease. There are many ways in which desirable control of high populations of rodents and insects or excessive abundance of noxious plants or plant diseases may be accomplished. A general

---

Plate 30

top. Wisconsin farmland on which the vegetative cover was so badly misused that a gully had carved its way well up the valley through productive soil. bottom. One year later the gully was dammed and converted to a useful pond as part of a complete conservation treatment of the watershed. The water is piped below the dam to supply livestock, fish are raised in the pond, and ducks and muskrats live in the grassy border.

outline of control methods, arranged roughly in order of the cost of application, follows:

1. Cultural
    Suitable tillage methods
    Regulated grazing
    Clearing and mowing
    Burning
    Flooding
2. Biological
    Change of habitat
    Encouragement of natural enemies and diseases
    Living barriers or buffers
3. Mechanical
    Proper construction
        Dams, terraces, etc.
    Non-living barriers
        Fences and other exclosures
    Trapping and shooting
4. Chemical
    Repellents
    Sprays, gases, poisons

It cannot be too strongly emphasized that permanently successful measures for the control of plants and animals adversely affecting operations on the land are, for the most part, methods of managing land rather than means of direct combat. They can, in most instances, become an accompaniment of proper land use, and when control becomes a result of management it is accomplished at a minimum cost. To this extent, control of harmful organisms remains primarily a responsibility of the land operator, whose greatest need is technical assistance with correct methods of operation. Resort to direct, aggressive control is frequently an unconscious admission that the land has been mismanaged or its use has been unsuccessful, and the remedy then lies far less in public support of control measures than in revised and improved methods of operation.

In spite of attempts to use land wisely, there remain conditions under which direct control may be desirable, especially until more is learned about the relation of injurious organisms to the management of land. The need for control may vary from year to year, as do the grasshopper hordes that attack crop fields in the Middle West, or it may depend upon seasonal or local conditions.

The control of injurious rodents, insects, plant diseases, and noxious plants should be related, biologically and economically, to the benefits to be derived from their control. With respect to land-management practices, control should depend primarily upon the way in which it tends to improve the efficiency of those practices. Control should not be proposed for its own sake, but should show a profit in terms of protection to or improvement of a land-management measure. With this benefit to land and its use in mind, control should result in advantages that are:

1. *Biologic.* Every effort should be made to determine whether the proposed method of control of a plant or animal species is likely to result in actual decrease of that organism and whether or not it will cause an increase of any other species as injurious or more harmful than the one at which control is directed. For example, wholesale control of coyotes, as already pointed out, can result in an undesirable increase of jack rabbits, which are even more difficult to control. Neither should control of a plant or animal be undertaken if it is likely to result in damage to a land-management practice, either the one it is intended to protect or another. For instance, chemical destruction of ditchbank weeds may actually cause failure of the ditch by filling it with erosional debris. In so far as possible, therefore, the effects of the control of a species should be determined before operations are undertaken.

2. *Economic.* Control of a plant or animal species should be undertaken only when the cost of control is commensurate with the advantages gained. Control which costs more than the value

of the land on which it is conducted may in some instances be warranted: for example, if it is to reduce newly introduced pests, like the white-fringed beetle which threatens serious damage to valuable crops. Usually, however, if regular control costs more than the value of the land, or of the increased profits to be derived from the land as a result of the control, it is unjustifiable. Furthermore, the cost of control must be reckoned not on initial application alone, but also on the cost of maintenance wherever that is necessary. For example, on rangeland the cost of rodent control must be considered over a period of years, for the initial poisoning is seldom biologically successful, periodic re-poisoning being required as a maintenance measure.

Wherever control is deemed necessary, it should be done at a time when it is most effective and should be followed by such cultural and management practices as will tend to continue the desirable effects of the control measures applied. The work should be planned and executed only after full consideration has been given to all the needs of the land, and should be accompanied by adequate application upon contiguous lands so that its effect may be reasonably widespread.

PREDATOR CONTROL

Of recent years many students, the game managers particularly, have dealt with predator-prey relationships. It has been argued that pure biological reasoning will not solve economic problems involved in loss to a farmer or rancher who is dependent upon domesticated animals that have been attacked by predatory birds or mammals. This is quite true, but the conclusion that broad-scale predator control is justified on the chance that it may prevent possible loss to an individual operator is untenable. Proper management methods can control wild animals' injuring crops or range livestock as successfully as they maintain high populations of game species. The latter is now being done without programs of predator control—the former should be attempted in a comparable way.

Ecological aspects of predation and predator control have already been discussed at some length in preceding chapters.

RODENT CONTROL

Whenever rodent control for the protection of crops, orchards, tree plantings, and structures such as dams and terraces is necessary, technical suggestions from the land-management biologist will help to assure efficient results. Control can be obviated if original design of structures includes attention to potential damage from burrowing rodents. For example, broad-base terraces that are tilled are less frequently damaged by rodents than narrow ones that are not cultivated. A dam may likewise possess a biological factor of safety if it has sufficient freeboard (height of dam above the spillway) and is thick enough to prevent penetration of burrows into the line of saturation, as stated previously.

Abandoned fields reseeded to range grasses as well as reseeded open range may require protection for a time from certain species of rodents, but subsequent control should be carefully weighed against the permanent advantages gained. Rodent control on open rangeland is frequently unsound from both biologic and economic standpoints and public support may well depend upon an answer to the question: Will the cost of rodent control over a period of years pay for itself in pounds of livestock produced as a result of the control?

The English animal ecologist, Martin A. C. Hinton (1932), has written well of a rodent problem many of us who live in modern homes believe to belong to the Dark Ages. When, however, we learn that in the United States annual losses caused by rats to food, grain, fabrics, poultry, and other products are now estimated to amount to nearly $200,000,000, we realize otherwise. The house or black rat, the common, Norway or brown rat, and the house mouse have been in Europe for a long time. The ancient Greeks trained the marten as a house cat. Rats and mice have traveled all over the world with man. Hinton points out that the black rat was the first to arrive in Great Britain, probably about the time of the Norman Conquest or the First Crusade. It was the animal primarily responsible for the spread of bubonic plague, the black death, which ravaged Europe in the Middle Ages, and for the English plague of 1665-6. For hundreds of years it was the only rat living in English houses and ships.

By the eighteenth century there arrived in England the brown rat, hardy migrant from the open plains of central Asia (the black rat stems from the forests of India and the Malay Peninsula). Whereas the black rat could not survive a single winter without the protection of human shelter, the brown rat spread all over the country, using rivers, canals, and drains to enter basements, walls, and garrets of buildings, driving the black rat before it. By the end of the nineteenth century, the black rat persisted only on ships and top floors of solidly built granaries and breweries in large ports.

In the early part of the present century, the English began to rebuild central London and great edifices were constructed 'in which the seeds of bankruptcy and traffic congestion could be sown with sufficient dignity.' There was an effort to make the basements and ground floors rat-proof, and the brown rat was being shut out, at least with moderate success. But here is the rub. Modern London has attractive, open skylights, the roofs of buildings are joined, and the streets are bridged with a network of wires and cables. As Hinton states, 'no primitively arboreal species could imagine a nearer approach to paradise,' and the black rat is once more the common rat of many parts of the city, so much so that London can become again in peril of the plague. Improved sanitation probably has less to do with protection than the fact that so many people move out of the city to the suburbs at night. A well-directed campaign to rid the city of rats plus a means of protecting the roofs can remove the threat of the pestilence. Among several biological principles involved in this story, Hinton concludes, is the dictum that 'if you create a vacancy, it is your own very difficult business to keep it vacant.'

CONTROL

Sprays, dusts, and other chemical applications become routine operations for the successful growth of many crop plants. Their use may involve complicated ecological interrelationships, however, as shown by recent insect-control work in Utah. Sweetened, poisoned bran used to control grasshoppers was believed to have caused a heavy loss of honey bees. Utah beemen were convinced that the poison material, in use only a few years, had reduced state honey production 50 per cent. Orchardists complained that the loss of bees reduced pollination of fruit trees. Soil conservationists believed that lessened seed production of clovers and other legumes useful for soil conservation also resulted.

Much insect control is similar to rodent control in that it can be accomplished most successfully by the land operator as a part of his regular work. Cultural methods such as use of appropriate rotations and time of planting are effective in preventing insect damage. The protection of fencerows from fire, in order to establish perennial plants, is very desirable, for regular burning perpetuates the annual weeds with which many harmful insects are normally associated. However, the control of some insects that occur in unusual numbers, such as grasshoppers, may be dealt with effectively only by special endeavor, and in such cases every effort should be made to assure control methods biologically most effective, under proper technical direction.

NOXIOUS PLANT CONTROL

Along roadsides, irrigation ditch banks, strip-mined lands, and many other areas, the replacement of weeds by useful plants is a definite contribution to weed control. For example, establishment of the perennial *Lespedeza sericea* prevents the invasion of undesirable plants along roadsides as it does on unproductive borders between woods and crop fields. Pavlychenko (1942) expresses this idea when he states that 'it is a sound principle of weed control to take advantage, wherever possible, of the relative competitive efficiency of weeds and crop plants.' In his Canadian experiments, Pavlychenko demonstrated that crested wheatgrass, sowed in leafy spurge (*Euphorbia esula*), nearly eliminated the spurge in four years, as it did quack grass and Canada thistle, all of them virile, pestiferous plants. In three years, crested wheat smothered thick stands of the equally persistent weeds, perennial sow thistle and toad flax. As the author recommends, 'Whenever it is possible to control a weed by making it compete with a crop which is more aggressive than the weed and highly adapted to the habitat where the weed is found, such a method of control or eradication is usually not only the soundest procedure but the most economical as well.'

Many noxious plants are most profitably controlled by proper cultural methods. A very interesting way in which such methods can influence weeds and production is shown by changes of land use in the Palouse-like wheat country of west-central Idaho. On the rolling hills there, it had been the practice to grow only wheat. This was done by fallowing the ground every other year, for there is not sufficient moisture in the soil to support wheat annually, and fallowing reduces the heavy stand of weedy plants that would otherwise develop and rob the soil of stored water. Under such use, erosion was taking a heavy toll of topsoil, and yields were declining. Weeds, such as Canada thistle, Russian thistle, morning glory, wild oats, and fanweed, were becoming yearly more prevalent. To improve conditions, a 6-year crop rotation was established. For two years the land is planted to a grass-legume mixture, usually mountain brome and sweet clover, followed by a year of peas. Then a year of wheat is grown, followed by another year of peas, or fallow, and finally a second year of wheat. The rotation is then repeated. Not only is erosion retarded remarkably by such a scheme, but the yield of wheat in two years equals that of three years under a wheat-fallow rotation. The pea crop provides clear profit, and the weeds so troublesome under the old method have nearly disappeared under the longer rotation system.

Where their growth and seeding habits cannot be controlled

by farm operations or changes in land use, noxious plants may constitute a real menace, and if their continued presence is the cause of diminished returns from the land, special control may be advisable, such as chemical or mechanical means of extermination. When these are undertaken, however, as with the destruction of cactus, mesquite, and juniper, the cost of continued control must be evaluated in terms of land products. Such control should be contingent upon its integration with sound land use and should be undertaken with due attention to (1) best time of year for control, (2) cultural practices likely to maintain the controlled condition, (3) use of crops or forage plants likely to prevent growth of noxious plants, and (4) adequate control on contiguous lands.

PLANT DISEASE CONTROL

The control of plant diseases is subject to much the same criteria as that for noxious plants. Except in the case of damaging diseases of extensive proportion, as white pine blister rust, control is an obligation most efficiently fulfilled by the land operator. In case of widespread diseases, control should be advised only after biologic and economic influences have been carefully considered, and the value of resulting benefits has been weighed against the cost of control.

'VERMIN' AND BOUNTIES

For some time in the United States, 'vermin' campaigns have been conducted, even under the auspices of enlightened state governments. Let us look at an example. A county collection of 'vermin,' for which an eastern state made bounty payments, recently fell into the hands of M. Graham Netting, an authority on reptiles and amphibians. He found one-third of the snakes in the collection to be pilot blacksnakes, which are known to eat small rodents. It has been estimated that a pilot blacksnake eats about as many field mice annually as inhabit an acre of alfalfa—

probably 100. On the conservative estimate that a field mouse can do 25 cents' worth of damage each year to an alfalfa field, Netting (1939) concludes:

The same relative proportion [of pilot blacksnakes] applied to the 79,481 snakes killed in 26 counties in the fiscal year 1933-34 results in the amazing figure of 26,493 pilot blacksnakes destroyed. On this assumption, the sponsors of the 'vermin' contests, in dooming so many mouse-catchers, enabled 2,649,300 destructive mice to continue their prolific existences. On a monetary basis, using the minimum evaluation of five dollars on each pilot, the loss in rodent damage, in a single year, to the agriculturists . . . may have been over $132,000 as a result of the slaughter of pilot blacksnakes alone.

Wholesale attempts to destroy reputedly injurious animals have long been made, and the payment of bounties for encouragement is likewise a practice of long standing. In the United States, the bounty system was inaugurated by the Massachusetts Bay Colony in 1630 by payment of a few pence for each timber wolf destroyed. The bounty system was an outcome of that era which tried to produce wildlife by elimination of whatever interfered with production. Bounty payments are even now being made in some parts of this country for weasels, hawks, owls, and other predators, largely encouraged by groups who still believe that predators decrease the supply of game species desired.

As we know from our observations on food chains and pyramid of numbers, there is scarcely a species which is not in some sense a predator upon another. To eliminate one member of the chain may have far-reaching and unexpected effects upon other members. Unless these effects are understood, it is folly to encourage destruction of the member it is *apparently* advantageous to eliminate. Hunters some years ago in Pennsylvania complained about the increase of poisonous snakes. It startled them to learn that perhaps the random killing of hawks and owls, then encouraged by some persons, might be the cause. Did not the destruction of these predatory birds permit an increase of mice and other small mammals, an abundance of which could serve as food for a large number of snakes?

REPELLENTS

In order to protect planting stock, orchards, and ornamental trees and shrubs from various rodents, rabbits, deer, and other mammals, numerous mixtures of malodorous, sticky substances have been prepared. These are usually applied to the plants in the hope that mammals that feed upon them will be repelled and the plants thus remain untouched. For the most part, such materials have been unsuccessful. Either what is repellent to the maker of such substances proves to be neither repulsive nor even unpalatable to the destructive animal, or the material does not remain in an offensive condition long enough to accomplish its intended purpose.

Some little success has been obtained by using substances which create fear in the animals to be repelled. Thus the use of n-butyl mercaptan, with an odor similar to that of predators like the skunk and weasel, kept rats away from stored grain. Honey, peculiarly enough, was found to hold the odor for about six months, longer than any other substance. Further study of such 'fear-producing' repellents may prove worth while, and provide worthy successors to many current repellents, which are often like a witch's brew of all manner of things ugly and repulsive to human touch and nostrils.

Although repellent substances might be useful if successful ones can be developed, the ecological approach to damage of this sort is likely to be the one most productive of permanent results, and a complement of proper land management rather than a 'control' measure.

LAND AND MAN

THAT there is a very close connection between the health of people and the land has long been suspected, although even today we have scant information about this relationship. From the nutritionists we hear a great deal about minerals and vitamins along with the carbohydrates, fats, and proteins we learned about when we had our readin', writin', and 'rithmetic. Calcium and phosphorus, iron and copper, iodine and fluorine, and a whole array of 'trace elements,' some needed in such small amounts as to defy analysis by ordinary chemical methods, are now known to be essential to human welfare, as, of course, they have always been.

But the obvious fact that we depend for these elements upon the soil has not been so widely heralded. To be sure, we obtain these nutrients from the meats, fish, fowl, eggs, milk, vegetables, fruits, and other things we eat, but the plants we consume and those that are eaten by the animals which serve us as food find essential elements in the soil in which they grow. If those elements are not in the soil, we cannot get them by eating the plants grown on the land or the animals that graze there. Spinach is not always the same, nor milk invariably of high nutritive value. The men who make cheese will tell you that the cheeses of one country are not like those of another, or of one region like those of some distant region. Differences of processing and of the moulds that play such an important role in cheese making are not so important as differences in the milk itself. And the differences in milk are due as much to the land as to the breed of livestock. The kind of pasture the cows graze and the kind of soil upon which the pasture plants grow are determining factors.

Soils vary considerably from place to place, and the variety of soil types on even a hundred-acre farm is in many regions astonishing. Yet, as we learned from a brief treatment of erosion in Chapter VII, it is the topsoil, more than any other portion of the soil profile, which is richest in available plant foods. Subsoil is less richly supplied with phosphates, potassium, calcium, nitrogen, and other important elements in available form than is the original surface soil, to say nothing of the subsoil's deficiency in texture, biota, and other desirable properties. Land-management methods that permit soil erosion rob us of highly important nutrient elements.

We have a habit of believing, because we can grow annual lespedezas, kudzu, certain grasses, as orchard grass and redtop, and other plants adapted to depleted soils on our millions of areas of seriously eroded lands, that we have not lost much. We may indeed have a cover, but the yield of crops grown on depleted soils is low and of poor quality, and cattle that feed on plants deficient in important elements are weakened, and the people dependent on the products of such land are undernourished.

The poor share-cropper of the Appalachian Piedmont or the deep South, who seldom stays on one farm more than a couple of seasons, and then loads his pitifully few worldly possessions on a shaky wagon and moves on, is an ultimate product of land that is poor. The 'Oakie' and 'Arkie' are not so much the results of poor human inheritance or the unscrupulous bankers as they are human symptoms of wasted land and false land management. (Plate 31, p. 226.)

The well-known dietary deficiency diseases, such as scurvy, rickets, pellagra, and beriberi, can be due as much to lack of essential elements in food as to lack of eating the proper foods. Many of us think we need not worry much about these diseases with modern means of production and transportation making available to many persons a diet highly varied and supposedly sufficient. Yet every year in the United States there are still 100,000 cases of pellagra.

The effects of soil upon health have also been listed by the dentists, who claim that accelerated erosion in a local area can decrease calcium sufficiently to cause increased dental caries among people dependent largely upon the land of that area for food. In the past, particularly, serious bone, skin, digestive, and nervous diseases, as well as other maladies, occurred in certain localities. Now we know, as Auchter (1939) recently pointed out, that 'many of these troubles resulted from restricted diets or from eating plant and animal products produced on soils either deficient in certain elements or containing elements injurious to health. Even today there are regions where such troubles as goiter, skin diseases, weak and deformed leg bones, mottled and furrowed teeth and nervous disorders are all too common.'

In his book on *Rats, Lice, and History*, Zinsser (1937) shows in a very readable way how animals influence human welfare and the very course of human history. The relation between rodents and plague is now well understood, and we know that wild rodents, as well as domesticated rats, harbor and spread the disease by serving as hosts to plague-carrying fleas. Recently a bird, the burrowing owl, which frequents ground squirrel burrows, was found to harbor sticktight fleas infected with the plague organism (Wheeler *et al.* 1941). Here is an instance where range management can aid human health, for if the number of livestock is so balanced with the carrying capacity of the range that grass cover is good and wild rodents reasonably few in numbers, there will be fewer reservoirs for disease-bearing fleas.

Not only are diseases thus frequently associated with the soil and the way we use it, but other human ills may hark back to the land. Tenancy, tax delinquency, and the settlement of sub-marginal areas on which good people become poor, disheartened, and dependent, are social maladies that draw much of their sustenance from the failure of man to relate himself properly to his environment. Here are fundamental relationships that must be interpreted and acted upon. The land-management biologist has a responsibility as basically related to social welfare as the politician, busi-

[224]

ness man, banker, economist, or anyone else whose task is to lend his particular skill and knowledge to the maintenance of a wholesome community.

There are many instances of minor importance to show that man's welfare is related to the land. The relation of land-management practices to biological phenomena recently has captured the imagination of many who are not practically concerned with land-use operations. For example, it has been claimed (Wodehouse 1940) that a cure for hay fever is soil conservation, for when idle and eroding lands are clothed with appropriate erosion-control vegetation there will be little place for ragweed to persist. The relation of bee-keeping to good land use has also been pointed out (Paddock 1937), for good seed and fruit production of many crop legumes and wildlife food plants useful in controlling erosion is dependent essentially upon pollination by bees.

LAND AND CULTURE

Although there is an important relation between the land and man's personal welfare, we must not make the mistake of believing that man is wholly a creature of his environment. Much in this book may suggest a cold, restrictive relation between natural conditions and desirable treatment of the land; it may connote an irrefutable dependence of man upon the habitat he occupies. Very much the contrary is actually intended. That man can influence his environment is implied in the assumption that he can learn how to manage the land (Plate 32).

A great deal has been written and said about culture as a reflection of environmental factors. We hear that the culture of a people grows out of, and is dependent upon, the physiography, soil, vegetation, and other factors we have considered of importance in land use. We have seen that climate determines a desert in one place and a rich grassland in another, that soils depend upon the interaction of several habitat factors, and that vegetation can be relied upon as a living expression of the total environment. Likewise animals, so dependent upon vegetation, display a remarkable correlation between their occurrence and numbers and the conditions under which they live. Man, as an animal, shows the effects of habitat upon his daily life and, to an extent, the culture he exhibits can be correlated with regional characteristics. Nomads are associated with open desert grassland, Eskimos with frozen wastes, and so with all the peoples and the lands they occupy as we learned very early from our old school geographies.

The relation between the desires and needs of man and the things in nature which he can utilize is not a simple one, and there is a difference between man and other animals which no manager —land, economic, or political—can afford to ignore. Between man and his environment there is interjected the power to reason. In spite of our habitual ways of doing things, every once in a while someone gets an *idea*. Although there is nothing utilitarian about an idea in itself, it may govern not only the food we eat but our dress also, and the construction of our dwellings. In fact a great deal of our physical and material well-being depends upon our ingenuity of thought. Not only ecologists, but politicians who lose sight of this fact let slip from their grasp one of the most potent influences in human progress.

The English anthropologist, C. D. Forde (1934), has illustrated how important ideas are, or how significant the lack of them can be. He studied a number of peoples who live outside the sphere of modern civilization in order to learn something of the relation between the human habitat and the manifold technical and social devices developed for its exploitation. Forde recognized the essential economies of collecting, hunting, fishing, cultivating, and

---

PLATE 31

TOP. On poor land the people are poor. The rundown house is no less an expression of man's failure to adapt himself to the land than the misused, eroded hillside, which should never have been cultivated.
BOTTOM. Where the land is cared for the homes are good and the people are prosperous. Unless man adapts himself and his ways to the capacity of the land to produce under careful management, his future is jeopardized.

stock rearing, and examined habits of primitive groups representing these economies in various parts of the world. He was cautious
to acknowledge, however, that some of these activities may be
practiced concomitantly and that the stages are not always distinct. The modern city dweller, for instance, still gathers blackberries for food and hunts game for pleasure and social prestige.

Tools, Forde reminds us, are the results of ideas, and they are
easily transmitted from one people to another. Thus in prehistoric
time, the dog sledge and tailored fur clothing spread throughout
all the human groups in the Arctic, from Lapland to Greenland.
The use of the horse as a beast of burden and the art of riding
spread throughout many parts of the Old World, and some people, like the Bedouins, even retain the horse under environmental
conditions to which it is not well adapted. In the New World,
the use of pottery spread throughout the Americas and the cultivation of corn also received widespread attention in prehistoric
time. The ancient Peruvians and other Americans developed
elaborate and extensive irrigation systems and made widespread
use of manures as fertilizers.

On the other hand, a civilization frequently reflects man's
failure to get an idea. In all the New World and Oceania, no
one ever thought of the wheel, although rollers were used to move
heavy objects. The plow was unknown in the western hemisphere
and beasts of burden were not used except for the alpaca and
llama, in the Andean highlands, and the dog, employed principally

---

PLATE 32

That man can quickly improve the landscape is well shown by these two air
views of a small farming community in the Northeast. The upper, photographed in 1936, shows traditional farming with 'square' fields. The lower,
taken in 1939, shows strip-cropped fields laid to the contour of the land.
Many other changes, as conversion of steep slopes to pasture, vegetated outlets to safely dispose of runoff water, and adapted crop rotations, are not
apparent. Adopted through co-operative effort of land operators, such changes
prove that man can consciously adapt his actions and direct his progress in
keeping with the demands of environmental factors.

[227]

as a draught animal by the Eskimos and the Indians of the Plains, British Columbia, and Alaska. Nor were the arts of milking and riding known here. Some people even refuse, for economic or other reasons, to use tools that are readily available to them. In southern China, for example, intensive and highly productive agriculture is practiced without the use of the plow, which has long been known there.

Thus it appears that man is independent of his environment to the extent that he can modify it. Cultures are not simply functions of habitat conditions or available natural resources, but are the products of the long accumulation and integration of customs, notions, and consciously directed activities. Forde concludes that the 'mere fact that there should be such a wide range of variation in the character and development of craft, economy, and society —between not only diverse habitats but those very similar in their general conditions—makes it obvious that much has depended on particular discoveries in particular areas, and on special trends of development in particular societies.'

All of this is by way of saying that while we must recognize the intimate connection between living things and the land, ideas of managing land can so influence conditions that they will be satisfactory to health and permanent security or detrimental to well-being and ultimate destiny, depending upon the thoughtfulness we display.

LAND AND SOCIETY

It is so easy to say what the world needs, that we must be wary of dogmatic practicalisms. Nevertheless, this book cannot close without a word about the value of principles, in our case natural ones, not only to the management of land, our primary concern here, but to society as well. H. G. Wells, on a trip to the United States in 1942, wrote an article for the press on the state of the war, in which he claimed that 'we need a type of scientific worker who does not yet exist. We need—what shall I call them?—professors of human ecology, or, if you want a less precise and pompous name, of foresight. It has even been suggested there

[228]

should be a "ministry of foresight." ' Certainly one of the most significant contributions of the land-management biologist is the prediction of consequences resulting from a knowledge of natural principles. Will the results prove unwise or redound to the preservation of land resources and the welfare of those who depend upon the land?

The lesson of total war has begun to have its effect upon other phases of national life. Total war means a war in which all the myriad activities of a complex society are co-ordinated, integrated, and directed toward a common objective—victory. Now we hear of total conservation and of a total peace in which our efforts are co-ordinated for the attainment of constructive goals. It is becoming clear in this world of expert knowledge and specialized technique that we need a purpose for knowledge and an amazing amount of synthesis. We need, in fact, co-ordination within an individual's thinking. That means an appreciation of relationships, not disturbed or diverted by knowledge of a great many facts or familiarity with a maze of details. As some of the social scientists have said, 'Knowledge which the sophisticated experts possess is growing at a rate more rapid than the rate at which it is being institutionalized in the habits of thought and action of the mass of our population.' (Lynd 1939 p. 108.)

Without a broad outlook and an attempt to relate knowledge to human welfare, the applied biologist or any other technician cannot be fully successful. Those who train ecologists have emphasized this responsibility. Paul Weiss (1942), in discussing the education of biologists, claims that, in the world ahead, their fitness will depend only partly on biological competence, and 'unless they are also made conscious of their obligation to society, they will . . . be at a loss to justify their subsidized existence to a society asking uncomfortable questions . . . And the biologist will have to be convincing in showing cause why what he is doing should not be discontinued as a publicly supported enterprise.'

Although we need to know a great many things, of 'shoes and ships and sealing wax and cabbages and kings,' knowledge for the

[229]

land manager must be related to the tasks at hand. Much ecological research has been done, but at every turn the land technician meets biological questions for which the answers are nowhere to be found. For research bearing upon problems of land management, there must always be support, as we support research for the making of industrial products, or methods of controlling insects injurious to crops. Management problems are complex problems, challenging the ingenuity of the researcher, and for their solution there is real need.

A pertinent story near the beginning of a recent book on a phase of our social structure (Allport 1933) is appropriate near the end of this one:

At a meeting of the faculty of a certain large university a proposal for a new administrative policy was being discussed. The debate was long and intense before a final vote of adoption was taken. As the professors filed out of the room an instructor continued the discussion with one of the older deans.

'Well,' observed the latter official, 'it may be a little hard on some people; but I feel sure that, in the long run, the new plan will be for the best interests of the institution.'

'Do you mean that it will be good for the students?' inquired the younger man.

'No,' the dean replied, 'I mean it will be for the good of the whole institution.'

'Oh, you mean that it will benefit the faculty as well as the students?'

'No,' said the dean, a little annoyed, 'I don't mean *that*; I mean it will be a good thing for the institution itself.'

'Perhaps you mean the trustees then—or the Chancellor?'

'No, I mean the institution, the *institution!* Young man, don't you know what an institution is?'

To many people, ecology and land management are very general terms, and they are frequently as loosely used as the dean used the concept of the institution. Like the equally general concept conservation, however, they are terms used in attempts to define movements of social significance still none too clear to us, yet growing and powerful forces. Land-management biology, as we have treated the subject in this book, has a part to play in conservation. And conservation is more than wise use and preservation of natural resources—mineral, soil and water, plant and animal, human. It may well be the foundation of a new social philosophy. Our public land policies, anti-trust regulations, tariffs, and related laws were developed around the exploitation of natural resources and they helped materially to promote our present economy. Is it not reasonable to suppose, then, that laws and regulations which promote the conservation of resources will help to build a new attitude toward the environment in which we live? And from this may well develop a way of life as different from the prevailing one as that is different from cultures of the past.

BIBLIOGRAPHY

THE following list consists largely of the sources cited in this book to illustrate natural principles. For this purpose, other references might have been used to as good advantage, although considerable care has been used in selecting examples. Such a list, in any event, can only be suggestive of the many appropriate sources of specific information available. In order to make the bibliography more generally useful to the land-management biologist, recommended standard works have been added and many of the references have been briefly annotated.

ABLEITER, J. K.
1937. 'Productivity Ratings in the Soil Survey Report.' *Soil Sci. Soc. Proc.* 16:415-22. Describes the productivity rating or productivity index, which is one of the numerous schemes for classifying land.

ACKERMANN, EDWARD A.
1941. 'The Köppen Classification of Climates in North America.' *Geog. Rev.* 31:105-11. A short but significant contribution on the way in which Köppen's concepts of climate can be applied to this continent.

ADAMS, CHARLES C.
1908. 'The Ecological Succession of Birds.' *The Auk* 25:109-53. One of the early articles on animal succession by an ecologist whose work is of interest to the land manager.

————
1913. *Guide to the Study of Animal Ecology.* Macmillan. 183 pp. Consists of an excellently annotated bibliography of important ecological work and a stimulating discussion of the dynamic point of view.

————
1935. 'The Relation of General Ecology to Human Ecology.' *Ecology* 16:316-35. One of the few good articles on this broad subject.

ADAMSON, R. S.
1939. 'The Classification of Life-forms of Plants.' *Bot. Rev.* 5:546-61. A comprehensive review and bibliography on the subject.

[233]

BIBLIOGRAPHY

ALLAN, PHILIP F.

1942. 'Defensive Control of Rodents and Rabbits.' *Jour. Wildl. Mangt.* 6:122-32. A good discussion of the naturalistic approach to the control problem.

ALLAN, PHILIP F., and CECIL N. DAVIS

1941. *Ponds for Wildlife.* U. S. Dept. Agr. Farm. Bul. 1879. 46 pp. Emphasizes the management of ponds as part of a soil-conservation program, especially the use of aquatic vegetation.

ALLEE, W. C.

1938. *The Social Life of Animals.* W. W. Norton. 293 pp. Elaborates the thesis that reproduction and the *per capita* survival of animals may be higher when many rather than few individuals are present. Thus undercrowding may be as important as overcrowding.

ALLEN, JOEL A.

1877. 'The Influence of Physical Conditions in the Genesis of Species.' *Radical Review* 1:108-40. The original statement of Allen's Rule.

ALLPORT, FLOYD H.

1933. *Institutional Behavior: Essays toward a Reinterpreting of Contemporary Social Organization.* Univ. of N. C. Press. 526 pp.

ANNAND, P. N.

1940. 'Recent Changes in Agriculture and their Effect on Insect Problems.' *Jour. Econ. Ent.* 33:493-8. Discusses relation of strip cropping, land conversions, and other conservation measures with respect to insects, but stresses need for more knowledge and co-ordinated effort.

ASHMAN, R. I., et al.

1936. 'Forest Wildlife Census Methods Applicable to New England Conditions.' *Jour. Forestry* 34:467-71.

ATLAS OF AMERICAN AGRICULTURE

1936. *Physical Basis, Including Land Relief, Climate, Soils, and Natural Vegetation of the United States.* Gov't Printing Office. Parts paged separately. An already classical treatment of the subjects listed, in folio size, with many colored maps. An unparalleled reference work for the land-management biologist.

ATWOOD, WALLACE W.

1940. *The Physiographic Provinces of North America.* Ginn & Co. 536 pp. Popular, readable book with folded, unusually good relief map of the United States, showing physiographic landforms.

AUCHTER, E. C.

1939. 'The Interrelation of Soils and Plant, Animal and Human Nutrition.' *Science* 89:421-7.

[234]

BIBLIOGRAPHY

BALDWIN, PAUL H., and GUNNAR O. FAGERLUND
1943. 'The Effect of Cattle Grazing on Koa Reproduction in Hawaii National Park.' *Ecology* 24:118-22. An example of animals maintaining in one type of vegetation—grassland—an area previously in another—forest.

BEECHER, WILLIAM J.
1942. *Nesting Birds and the Vegetation Substrate.* Chicago Ornith. Soc. 69 pp. An important paper supporting the contention that 'the population density of most nesting birds varies as a direct function of the amount of edge per unit area.'

BELL, W. B.
1937. 'Methods in Wildlife Censuses.' *Jour. Am. Statistical Assoc.* 32:537-42. A brief summary in general terms of methods for censusing birds and mammals.

BENNETT, H. H.
1921. 'The Classification of Forest and Farm Lands in the Southern States.' *South. Forestry Cong. Proc.* 3:69-113. One of the earliest papers calling for a systematic evaluation of land with respect to its adapted use.

——— 1932. 'Relation of Erosion to Vegetative Changes.' *Sci. Monthly* 35:385-415. Broad discussion of the occurrence and succession of plants on severely eroded soils in various parts of North America.

——— 1939. *Soil Conservation.* McGraw-Hill. 993 pp. A standard, comprehensive text on all aspects of the subject.

——— 1942. 'Total Conservation.' *Soil Conservation* 7:233-5. A short article emphasizing the importance of attending to the 'wildlife lands' of agricultural areas.

BENNETT, LOGAN J., and GEORGE O. HENDRICKSON
1939. 'Adaptability of Birds to Changed Environment.' *The Auk* 56:32-7. A study showing that marsh and prairie birds are no less abundant now in a northwestern Iowa area than they were thirty years ago when agriculture was less intensive.

BERCAW, LOUISE O., and ANNIE M. HANNAY
1938. *Bibliography on Land Utilization, 1918-36.* U. S. Dept. Agr. Misc. Publ. 284. 1508 pp. Comprises 7300 annotated references.

BIRD, JOHN A.
1942. *Western Ground Waters and Food Production.* U. S. Dept. Agr. Misc. Publ. 504. 40 pp. Describes a number of instances in which thoughtless use of land and water reduced the ground water supply.

[235]

BIBLIOGRAPHY

BISHOPP, F. C.
1938. 'Entomology in Relation to Conservation.' *Jour. Econ. Ent.* 31:1-11. Excellent remarks on the relation of insects to erosion, soil building, clean-culture, wildlife, forest and range management, and other aspects of conservation.

BOND, RICHARD M.
1939. 'Coyote Food Habits on the Lava Beds National Monument.' *Jour. Wildl. Mangt.* 3:180-98.

BOYCE, J. S.
1941. 'Exotic Trees and Disease.' *Jour. Forestry* 39:907-13. A case for native species.

BRASS, L. J.
1941. 'Stone Age Agriculture in New Guinea.' *Geog. Rev.* 31:555-69.

BRAUN-BLANQUET, J.
1932. *Plant Sociology*. McGraw-Hill. 439 pp. American edition of the German *Pflanzensoziologie*. Transl. by G. D. Fuller and H. S. Conard. It serves to acquaint us with the nomenclature of European plant ecology, and many of its concepts.

BUREAU OF ANIMAL POPULATION
1936-37. *Annual Report* 1936-37. Univ. of Oxford, England. 38 pp. The report includes the annual summary of the work of the Bureau and comments on the significance of animal numbers.

CAHN, A. R.
1938. 'A Climographic Analysis of the Problem of Introducing Three Exotic Game Birds into the Tennessee Valley and Vicinity.' *Trans. Third North Amer. Wildl. Conf.* 807-17.

CAIN, STANLEY A., and WM. T. PENFOUND
1938. 'Aceretum rubri: The Red Maple Swamp Forest of Central Long Island.' *Amer. Midl. Nat.* 19:390-416. An example of the use of the species-area curve for determining the size of sample plots necessary to interpret correctly the composition of vegetation.

CARPENTER, J. RICHARD
1938. *An Ecological Glossary*. Univ. Okla. Press. 306 pp. The only comprehensive compilation of ecological terms, with definitions and references.

CHAMBERLAIN, T. C., *et al.*
1877-83. *Geology of Wisconsin: Survey of 1873-1879.* 4 vol. and Atlas. Part II, Geol. of eastern Wisc.; Chapter III, Native Vegetation, pp. 176-87. 1877. 2nd ed. 1878. Publ. by the State. This early discussion of vegetation types includes recognition of plant cover as an indicator of agricultural use.

[236]

BIBLIOGRAPHY

CHANDLER, ROBERT F., JR.
1940. 'The Influence of Grazing upon Certain Soil and Climatic Conditions in Farm Woodlands.' *Jour. Amer. Soc. Agron.* 32:216-30. A study conducted in central New York.

CHAPMAN, ROYAL N.
1931. *Animal Ecology: With Especial Reference to Insects.* McGraw-Hill. 464 pp. A good general work on insect ecology, with replete bibliographies for each section.

CHASE, W. W., et al.
1942. 'The Effects of Forest Harvest on Game Production.' *Jour. Forestry* 40:639-41.

CLEMENTS, FREDERICK E.
1928. *Plant Succession and Indicators: A definitive edition of Plant Succession and Plant Indicators.* H. W. Wilson Co. 453 pp. This is a combined and condensed edition of *Plant Succession* and *Plant Indicators* published by the Carnegie Institution of Washington in 1916 and 1920 respectively. It presents in technical language much that is fundamental on the subject.

——— 1935. 'Experimental Ecology in the Public Service.' *Ecology* 16:342-63. A brief discussion of the value of ecological studies in land classification, and their contribution to a knowledge of erosion, management of the public domain, windbreaks, and natural landscaping.

——— 1938. 'Climatic Cycles and Human Populations in the Great Plains.' *Sci. Monthly* 47:193-210. Discusses rainfall cycles and their relation to settlement and land use in the Dust Bowl.

CLEMENTS, FREDERICK E., and V. E. SHELFORD
1939. *Bio-ecology.* John Wiley & Sons. 425 pp. A textual work attempting to treat plants and animals as interactive members of the biotic community. The needlegrass-antelope complex is treated in detail.

CLEMENTS, FREDERICK E., JOHN E. WEAVER, and HERBERT C. HANSON
1929. *Plant Competition: An Analysis of Community Functions.* Carn. Inst. Wash. Publ. 398. 340 pp. A summary work on the subject.

COILE, THEODORE S.
1940. *Soil Changes Associated with Loblolly Pine Succession on Abandoned Agricultural Land in the Piedmont Plateau.* Duke Univ., School For. Bul. 5. 85 pp.

COMPTON, LAWRENCE V.
1943. 'Techniques of Fishpond Management.' *U. S. Dept. Agr. Misc. Publ.* 528. 22 pp. A summary review of the subject, with emphasis on fertilizing pond waters.

BIBLIOGRAPHY

COMPTON, LAWRENCE V., and R. FRANK HEDGES
  1943. 'Kangaroo Rat Burrows in Earth Structures.' *Jour. Wildl. Mangt.* 7:306-16. Illustrates how a knowledge of mammal habits can influence engineering design to the extent that expensive mammal control may be obviated.

COX, W. T.
  1938. 'Snowshoe Hare Useful in Thinning Forest Stands.' *Jour. Forestry* 36:1107-9. A specific example of the beneficial value of a mammal in the Lake States forests.

CUYLER, ROBERT H.
  1931. 'Vegetation as an Indicator of Geologic Formations.' *Bul. Am. Assoc. Petroleum Geol.* 15:67-78. A study near Austin, Texas, which shows a correlation between vegetation types and underlying geological formations.

DALQUEST, W. W., and V. B. SCHEFFER
  1942. 'The Origin of the Mima Mounds of Western Washington.' *Jour. Geol.* 50:68-84. An interesting study describing how the work of pocket gophers influences forest succession. See also Larrison, 1942.

DAMBACH, CHARLES A.
  1942. 'Fence Row Facts.' *Soil Conservation* 7:238. Points out the beneficial insects, birds, and mammals which populate protected Ohio and Indiana fence rows.

———
  1944. 'A Comparative Study of the Productiveness of Adjacent Grazed and Ungrazed Sugar Maple Woods.' *Jour. Forestry* 42: In Press. An important study showing the effect of grazing upon soil condition and organisms, small mammals, tree growth, and maple-syrup production in an Ohio woodlot.

DAMBACH, CHARLES A., and E. E. GOOD
  1940. 'The Effect of Certain Land Use Practices on Populations of Breeding Birds in Southwestern Ohio.' *Jour. Wildl. Mangt.* 4:63-76. A significant paper on the relation of bird numbers to conservation practices such as strip cropping and the protection of woodlots from grazing.

DARLING, J. N.
  1934. 'Statement of Hon. Jay N. Darling, Chief, Bur. of Biol. Survey, Dept. Agriculture' (8 May 1934) in *Hearings before Spec. Comm. on Cons. of Wildl.*, House of Rep. 73rd Cong., pp. 1-18, 101-18, 153-8. Includes one of the early statements about the desirability of converting eroded areas into wildlife habitats.

DAUBENMIRE, REXFORD F.
  1938. 'Merriam's Life Zones of North America.' *Quart. Rev. Biol.* 13:327-32. A summary of the criticisms of Merriam's life-zone concept.

BIBLIOGRAPHY

DAVISON, VERNE E.

1939. *Protecting Field Borders.* U. S. Dept. Agr. Leaflet 188. 8 pp. Specifications on how and what to plant on the eroding edges of crop fields adjoining woodland.

———

1942. 'Thirty Million Acres of Undiscovered Wildlife Land in the United States.' *Trans. Seventh North Amer. Wildl. Conf.* 366-74. Deals with the recognition and treatment of those types of agricultural land adapted best to the production of wildlife.

DAVISON, VERNE E., and J. A. JOHNSON

1943. *Fish for Food from Farm Ponds.* U. S. Dept. Agr. Farm. Bul. 1938. 22 pp. Practical guide for constructing, stocking, and fertilizing fish ponds.

DICE, LEE R.

1930. 'Methods of Indicating Relative Abundance of Birds.' *The Auk* 47:22-4.

———

1931a. 'Methods of Indicating the Abundance of Mammals.' *Jour. Mammalogy* 12:376-81.

———

1931b. 'The Relation of Mammalian Distribution to Vegetation Types.' *Sci. Monthly* 33:312-17. Supports the thesis that the occurrence of mammals is related more to vegetation and soil types than to individual plant species.

———

1938a. 'Some Census Methods for Mammals.' *Jour. Wildl. Mangt.* 2:119-30.

———

1938b. 'Poison and Ecology.' *Bird Lore* 40:12-17. A plea that control of mammals be based less on aggressive methods and more on ecological knowledge of the animals and the biological complex of which they are a part.

———

1941. 'Methods for Estimating Populations of Mammals.' *Jour. Wildl. Mangt.* 5:398-407.

DODD, ALAN P.

1940. *The Biological Campaign against Prickly-pear.* Commonwealth Prickly Pear Board. Queensland, Australia. 177 pp. Report of an outstanding instance of biological control of a plant by an insect.

BIBLIOGRAPHY

DONALD, C. M.

1941. 'Land Use.' *Jour. Aust. Inst. Agric. Sci.* 7:96-104. Discusses the cropping districts of Australia, relation of natural factors to grazing use, and some general aspects of land management.

DREW, WILLIAM B.

1942. *The Revegetation of Abandoned Crop Land in the Cedar Creek Area, Boone and Calloway Counties, Missouri.* Mo. Agr. Expt. Sta. Res. Bul. 344. 52 pp. A good paper which summarizes much of the literature on old-field succession.

EDMINSTER, F. C.

1937. 'An Analysis of the Value of Refuges for Cyclic Game Species.' *Jour. Wildl. Mangt.* 1:37-41. A New York example of a case in which a refuge area did not support higher populations of certain vertebrates than an adjacent hunted area.

———

1942. 'More Ruffed Grouse from Woodland Management.' *Soil Conservation* 7:261-2. Presents management recommendations.

ELLIS, M. M.

1936. 'Erosion Silt as a Factor in Aquatic Environments.' *Ecology* 17:29-42. Emphasizes the detrimental influence of silt upon aquatic organisms and habitat.

———

1937a. 'Detection and Measurement of Stream Pollution.' *U. S. Dept. Commerce, Bur. Fisheries Bul.* 48:365-437.

———

1937b. 'Pollution and Aquatic Life.' *Trans. Second North Amer. Wildl. Conf.* 653-7. Points out deleterious effects of various stream pollutants.

ELTON, CHARLES

1936. *Animal Ecology.* Macmillan. 209 pp. A most readable and authoritative book dealing in large part with the dynamics of animal communities and animal numbers. The best of its kind.

———

1942. *Voles, Mice, and Lemmings: Problems in Population Dynamics.* Oxford Univ. Press. 496 pp. A comprehensive treatment of what has been learned at Oxford's Bureau of Animal Population about fluctuations in animal numbers.

ELY, RICHARD T., and GEORGE S. WEHRWEIN

1940. *Land Economics.* Macmillan. 512 pp.

EMERSON, G. B.

1850. *Report on the Trees and Shrubs Growing Naturally in the Forests of Massachusetts.* Dutton and Wentworth. 547 pp. One of the early American works in which plant succession was recognized.

BIBLIOGRAPHY

ENDERS, ROBERT KENDALL

1930. *Some Factors Influencing the Distribution of Mammals in Ohio.* Occ. Papers Mus. Zoo., Univ. Mich. No. 212. 27 pp. Deals primarily with physiographic features, and concludes that such factors do not explain the occurrence of mammals.

ERRINGTON, PAUL L., and F. N. HAMERSTROM, JR.

1936. 'The Northern Bob-whites' Winter Territory.' *Ia. Agr. Expt. Sta. Res. Bul.* 201. pp. 301-443. An outstanding work on predation.

FAWCETT, C. B.

1930. 'The Extent of the Cultivable Land.' *Geog. Jour.* 76:504-509. A world-wide consideration.

FENNEMAN, NEVIN M.

1931. *Physiography of Western United States.* McGraw-Hill. 534 pp. This and the following book are standard references. This one contains a folded outline map of the physiographic provinces of the entire United States.

———

1938. *Physiography of Eastern United States.* McGraw-Hill. 714 pp.

FENTON, F. A.

1940. 'Does Burning Control Insect Pests?' Unpublished mss. The author contends that it does not.

FERGUS, E. N.

1936. 'Shall Crops Be Adapted to Soils or Soils to Crops?' *Jour. Amer. Soc. Agron.* 28:443-6. A stimulating paper pointing out that raising lespedezas and certain grasses adapted to depleted soils may only serve to further damage those soils.

FINNELL, H. H.

1939. *Problem-area Groups of Land in the Southern Great Plains.* U. S. Dept. Agr., unnumbered publication. 40 pp., with folded maps. A good example of the problem-area type of land planning.

FITZPATRICK, FREDERICK L.

1940. *The Control of Organisms.* Teachers College, Columbia Univ. 334 pp. A general treatment of human diseases, plant and animal parasites, insects, rodents, weeds, and other organisms which compete with man.

FORBES, S. A.

1895. 'On Contagious Disease in the Chinch-bug (*Blissus leucopterus* Say).' *19th Rpt. Sta. Ent. Ill.* pp. 16-176. In this article the author stated that economic entomology was applied ecology, thus being one of the first to recognize ecology's practical significance.

FORDE, C. DARYLL

1934. *Habitat, Economy and Society: A Geographical Introduction to Ethnology.* Methuen & Co., London. 500 pp.

[241]

BIBLIOGRAPHY

FORMAN, JONATHAN

1943. 'Soil Conservation Can Help.' *Soil Conservation* 8:267-70, 274, 284. Supports the thesis that human nutrition depends upon the quality of food, which in turn depends upon the quality of the soil.

FORSTER, H. C.

1941. 'The Use of Climatic Graphs in Determining Suitable Areas for Flax Production.' *Jour. Dept. Agr. Victoria* (Melbourne) 39:515-24. One of the few examples of the practical application of the climograph to plant production.

GABRIELSON, I. N.

1939. 'The Correlation of Water Conservation and Wildlife Conservation.' *Ill. Cons.* 4:12-13. Emphasizes various values of water areas and the dangers of excessive drainage.

———

1942. *Wildlife Conservation.* Macmillan. 250 pp. A general, popular, and authoritative book, with some chapters on the relation of wildlife to other kinds of conservation.

———

1943. *Wildlife Refuges.* Macmillan. 257 pp. A comprehensive treatment of the wildlife refuges of North America. There is a bibliography.

GAINES, STANLEY H.

1938. *Bibliography on soil erosion and soil and water conservation.* U. S. Dept. Agr. Misc. Publ. 312. 651 pp. Comprises 4388 annotated references, largely from the decade preceding date of publication.

GLEASON, H. A.

1926. 'The Individualistic Concept of the Plant Association.' *Bul. Torr. Bot. Club* 53:7-26. Supports the thesis that the composition of plant communities may result as much from chance as from selection of habitat or other influences.

GOULD, E. W.

1935. 'Occurrence of Low-Growing Game Foods During the Oldfield White Pine-Mixed Hardwood Succession in the Harvard Forest.' Unpubl. ms.

GRAHAM, EDWARD H.

1940. 'Ecology and Land Use.' *Soil Conservation* 6:123-8.

———

1942. 'Soil Erosion as an Ecological Process.' *Sci. Monthly* 55:42-51.

———

1943. 'Land Classification as a Technique in Wildlife Management.' *Trans. Eighth North Amer. Wildl. Conf.* 360-69.

BIBLIOGRAPHY

GRAHAM, SAMUEL A.
1929. 'The Larch Sawfly as an Indicator of Mouse Abundance.' *Jour. Mammalogy* 10:189-96. Good example of the value of small mammals to forest welfare.

GRANGE, WALLACE B., and W. L. McATEE
1934. *Improving the Farm Environment for Wild Life.* U. S. Dept. Agr. Farm. Bul. 1719. 62 pp. One of the early articles stressing value of erosion-control plantings for wildlife.

GRAY, L. C., et al.
1924. 'The Utilization of our Lands for Crops, Pasture, and Forests.' *U. S. Dept. Agr. Year Book* 1923:415-506. Report of a committee recommending a systematic classification of the reserve land areas of the United States.

GRINNELL, JOSEPH
1924. 'Wild Animal Life as a Product and as a Necessity of National Forests.' *Jour. Forestry* 22:837-45. An important paper on the dependence of forests upon various kinds of animal life.

HACKETT, L. W., R. F. RUSSELL, J. W. SCHARFF, and R. SENIOR WHITE
1938. 'The Present Use of Naturalistic Measures in the Control of Malaria.' *Bul. of the Health Organization of the League of Nations* 7:1016-64. An excellent outline of the various methods of controlling an injurious insect, the mosquito.

HAECKEL, ERNST HEINRICH
1869. 'Ueber Entwickelungsgang und Aufgabe der Zoologie.' *Jenaische Ztschr.* 5:353-70. The paper in which the term ecology was first used and defined.

HALL, SIR A. D.
1941. *Reconstruction and the Land: An Approach to Farming in the National Interest.* Macmillan, 287 pp.

HALL, E. RAYMOND
1942. 'Gestation Period in the Fisher with Recommendations for the Animal's Protection in California.' *Calif. Fish and Game* 28:143-147. Describes the relationship between the fisher and several environmental factors.

HAMILTON, W. J., JR., and DAVID B. COOK
1940. 'Small Mammals and the Forest.' *Jour. Forestry* 38:468-73. A noteworthy article showing the value of small mammals to forests in New York.

HANSON, HERBERT C.
1939. 'Ecology in Agriculture.' *Ecology* 20:111-17. A good paper on ecological relationships, touching upon human welfare and land use.

[243]

BIBLIOGRAPHY

HARRIS, K. L.

1939. 'Soil Conservation Versus Insect Control.' *Ent. Soc. Wash. Proc.* 41:20-26. A brief discussion of conflicting cultural practices recommended by the entomologist and soil conservationist and the need for reconciling them.

HARVARD COLLEGE

1941. *The Harvard Forest Models.* Cornwall Press. 48 pp. Photographs and descriptions of several museum dioramas illustrating the history and use of land in central Massachusetts since primeval time.

HAWBECKER, A. C.

1940a. 'Planting for California Wildlife.' *Calif. Fish and Game* 26:271-7. A study of bird numbers as influenced by new woodland plantations.

————

1940b. 'The Burrowing and Feeding Habits of *Dipodomys venustus*.' *Jour. Mammalogy* 21:388-96. Emphasizes importance of annual plants as food for this kangaroo rat.

HENDERSON, W. C.

1930. 'The Control of the Coyote.' *Jour. Mammalogy* 11:336-53.

HENDRICKS, H. E.

1936. *A Land Use and Soil Management Program for Tennessee.* Univ. Tenn. Publ. 197. 24 pp. Includes a classification of rural land for the state.

HESKE, FRANZ

1938. *German Forestry.* Yale Univ. Press. 342 pp. A book which summarizes much of the experience with forest management in Germany.

HESSE, R., W. C. ALLEE, and K. P. SCHMIDT

1937. *Ecological Animal Geography.* John Wiley & Sons. 597 pp. A translation of Richard Hesse's *Tiergeographie auf oekologischer Grundlage,* 1924, a classical work on the relation of animal distribution to ecological facts and principles.

HILGARD, EUGENE W.

1858. *Report on the Geological and Agricultural Survey of the State of Mississippi.* Mississippian Steam Power Press Print, Jackson. 22 pp. Includes early recognition of the value of trees as indicators of adapted land use.

HINTON, MARTIN A. C.

1932. 'Biological Principles in the Control of Destructive Animals.' *Proc. Linnaean Soc.,* Session 144:111-26. Stresses the need for an ecological evaluation of undesirable animals, and caution about controlling them without first attempting carefully to regard the consequences.

[244]

BIBLIOGRAPHY

HOCKENSMITH, R. D., and J. G. STEELE
    1943. *Classifying Land for Conservation Farming*. U. S. Dept. Agr. Farm. Bul. 1853. 46 pp. A detailed description of the scheme for classifying rural land according to use capabilities, that is, the inherent capacity of the land to produce.

HOFMANN, J. V.
    1923. 'Furred Forest Planters.' *Sci. Monthly* 16:280-83. On the value of small mammals in the natural regeneration of forests.

HOLT, ERNEST G.
    1934. 'Erosion Control and Game Development.' *Trans. 20th Amer. Game Conf.*, 177-81. One of the early articles on the integration of soil and wildlife conservation by the person who subsequently became the foremost exponent of the subject.

HORT, SIR ARTHUR
    1916. *Theophrastus—Enquiry into Plants—and Minor Works on Odours and Weather Signs*. G. P. Putnam's Sons, Vol. 1, 475 pp., Vol. 2, 499 pp. An English rendition of the work of the Greek naturalist who lived three centuries B.C. and who treated many subjects essentially ecological.

HOWARD, H. ELIOT
    1920. *Territory in Bird Life*. Murray, London. 308 pp. The role of territory systems in the lives of English warblers.

HUBBS, CARL L.
    1933. 'Sewage Treatment and Fish Life.' *Sewage Works Jour.* 5:1033-40. A general discussion of the effects of sewage upon fish.

HUMBOLDT, ALEXANDER VON
    1805. *Essai sur la géographie des plantes; accompagné d'un tableau physique des régions équinoxiales, fondé sur des measures exécutées, depuis le dixième degré de latitude boréale jusqu'au dixième degré de latitude australe, pendant les années 1799, 1800, 1801, 1802, et 1803*. Par Al. de Humboldt, et A. Bonpland. Rédigé par Al. de Humboldt. Paris, Levrault, Schoell et compagnie. The first attempt to classify natural areas of the world.

HYSLOP, J. A., R. L. WEBSTER, and W. E. HINDS
    1925. 'Report of the Committee on Estimating Insect Abundance.' *Jour. Econ. Ent.* 18:24-32. A summary of census methods.

JACKS, G. V., and R. O. WHYTE
    1939. *The Rape of the Earth: A World Survey of Soil Erosion*. Faber & Faber, London. 313 pp. An excellent book on the subject, especially the general social and economic effects of erosion. A 332-page American edition was published by Doubleday, Doran in 1939 entitled *Vanishing Lands; a World Survey of Soil Erosion*.

BIBLIOGRAPHY

JACOT, ARTHUR PAUL

1936. 'Soil Structure and Soil Biology.' *Ecology* 17:359-79. Studies on the microfauna with respect to soil depth and use.

JAMES, PRESTON E.

1935. *An Outline of Geography.* Ginn & Co. 475 pp. A geography which is essentially ecological in its treatment of natural areas and their living components.

JESNESS, OSCAR B., et al.

1935. *A Program for Land Use in Northern Minnesota.* Univ. of Minn. Press. 338 pp. A type study in land utilization as applied to 14 northeastern Minnesota counties.

JUST, THEODOR (EDITOR)

1939. 'Plant and Animal Communities.' *Amer. Midl. Nat.* 21:1-255. Comprising the Proceedings of the Conference on Plant and Animal Communities, held at the Biological Laboratory, Cold Spring Harbor, Long Island, N. Y., 29 August to 2 September 1938. Ten papers by as many authors, dealing mostly with theoretical concepts and interpretations of living communities.

KEARNEY, T. H., et al.

1914. 'Indicator Significance of Vegetation in Tooele Valley, Utah.' *Jour. Agr. Research* 1:365-417. One of the early papers on the value of plants as indicators of agricultural use of land.

KELSO, LEON H.

1939. *Food Habits of Prairie Dogs.* U. S. Dept. Agr. Cir. 529. 16 pp.

KING, RALPH T.

1942. 'Is It Wise Policy to Introduce Exotic Game Birds?' *Audubon Magazine* 44:136-45, 230-36, 306-10. The author holds that it is generally unwise.

KLAGES, KARL H. W.

1942. *Ecological Crop Geography.* Macmillan. 615 pp. Primarily a treatment of factors influencing individual crops and a description of their worldwide distribution. Community ecology is not considered.

KOOPER, W. J. C.

1927. 'Sociological and Ecological Studies in the Tropical Weed-vegetation of Pasuruan (the Island of Java).' *Rec. Trav. Bot. Néerl.* 24:1-225. Shows that only a small percentage of the total weed flora acts as pioneer plants on disturbed areas.

KÖPPEN, W.

1931. *Grundriss der Klimakunde.* W. de Gruyter & Co., Berlin and Leipzig. 388 pp. 2nd, revised ed. of *Die Klimate der Erde.* A classical work on the most widely used classification of climates.

BIBLIOGRAPHY

KORSTIAN, C. F., and THEODORE S. COILE
    1938. *Plant Competition in Forest Stands*. Duke Univ. School Forestry Bul. 3. 125 pp. A study which 'demonstrates that competition between individuals of the forest vegetation for soil moisture is a highly significant factor in the growth, development, and reproduction of forests in the Piedmont plateau.'

KORSTIAN, C. F., and P. W. STICKEL
    1927. 'The Natural Replacement of Blight-killed Chestnut in the Hardwood Forests of the Northeast.' *Jour. Agr. Res.* 34:631-48.

LACK, DAVID
    1937. 'A Review of Bird Census Work and Bird Population Problems.' *Ibis* 50:369-95.

LA MONT, T. E.
    1937. *Land Utilization and Classification in New York*. Cornell Ext. Bul. 372. 60 pp. Describes the land classification used in the state and its relation to agriculture, roads, electric power, and reforestation.

LAND CLASSIFICATION, NATIONAL CONFERENCE ON
    1940. *The Classification of Land*. Proceedings of the First National Conference on Land Classification. Mo. Agr. Expt. Sta. Bul. 421. 334 pp. Numerous papers on most aspects of the subject.

LAND UTILIZATION, NATIONAL CONFERENCE ON
    1932. *Proceedings of the National Conference on Land Utilization*. Held in Chicago, Ill., 19-21 Nov. 1931. Gov't Printing Office. 251 pp. Numerous papers by various authors.

LANGLOIS, THOMAS H.
    1941. 'Two Processes Operating for the Reduction in Abundance or Elimination of Fish Species from Certain Types of Water Areas.' *Trans. Sixth North Amer. Wildl. Conf.*, 189-201. Discussion of the effect of erosion silt on fish habitat and the composition of fish communities.

LARRISON, EARL J.
    1942. 'Pocket Gophers and Ecological Succession in the Wenas Creek Region of Washington.' *The Murrelet* 23:35-41. Describes how pocket gophers, even on rocky lava plateaus, raise soil on which grasses and pines eventually grow. See also Dalquest and Scheffer, 1942.

LARSON, FLOYD
    1940. 'The Role of the Bison in Maintaining the Short Grass Plains.' *Ecology* 21:113-21. Holds that the grazing of bison determined the vegetation type existing before the advent of white men.

LAY, DANIEL W.
    1938. 'How Valuable are Woodland Clearings to Wildlife?' *Wilson Bul.* 50:254-6. Presents counts to show the edge effect displayed by birds, Walker County, Texas.

BIBLIOGRAPHY

LEOPOLD, ALDO

    1925. 'Wilderness as a Form of Land Use.' *Jour. Land and Public Utility Economics* 1:398-404. A plea for the establishment of wild roadless areas where one can enjoy primitive modes of travel and subsistence, that they might help preserve the American pioneer spirit and culture.

———

    1931. *Report on a Game Survey of the North Central States.* Sporting Arms and Ammunition Manufacturers' Institute. Madison, Wisconsin. 299 pp. Probably the first article to point out the feasibility of integrating soil and wildlife conservation.

———

    1933. *Game Management.* Charles Scribner's Sons. 481 pp. The standard text on wildlife management.

———

    1941. 'Cheat Takes Over.' *The Land* 1:1-4. Comments on cheat grass (*Bromus tectorum*) as a weed in western United States and on the responsibilities of the land manager.

LIEBIG, BARON JUSTUS VON

    1859. *Letters on Modern Agriculture.* Walton & Maberly, London. 284 pp. Contains the original statement of the idea now known as Liebig's Law of the Minimum.

LINCOLN, FREDERICK C.

    1930. *Calculating Waterfowl Abundance on the Basis of Banding Returns.* U. S. Dept. Agr. Cir. 118. 4 pp. Description of one of the most useful bird census methods.

LOVEJOY, P. S.

    1925. 'Theory and Practice in Land Classification.' *Jour. Land and Publ. Util. Econ.* 1:160-75. Primarily a discussion of the subject as it refers to Michigan, but with some more general comments.

LOWDERMILK, W. C.

    1939. 'Fish Ponds and Fields in Rotation.' *Soil Conservation* 5:123-6. Brief description of the practice followed in the Rhone Valley, France.

LYND, R. S.

    1939. *Knowledge for What?* Princeton Univ. Press. 268 pp. Deals with the place of social science in American culture.

MARCOVITCH, S.

    1935. 'Experimental Evidence on the Value of Strip Farming as a Method for the Natural Control of Injurious Insects with Special Reference to Plant Lice.' *Jour. Econ. Ent.* 28:62-70.

BIBLIOGRAPHY

MARSH, GEORGE P.
1864. *Man and Nature; or, Physical Geography as Modified by Human Action.* Scribners. 560 pp. Later published under the title, *The Earth as Modified by Human Action.* A forward-looking treatment of man's relation to the land, including a discussion of accelerated erosion due to deforested slopes.

MAYALL, K. M.
1938. *Natural Resources of King Township, Ontario.* Privately published, Toronto. Includes a comparison of bird numbers in grazed and ungrazed woods.

MCATEE, W. L.
1934. 'Food Habits of Predatory Mammals.' *Jour. Mammalogy* 15:243-4.

———
1936. *Wildlife Technology.* Wildlife Research and Management Leaflet BS-67. 5 pp. Mimeo. 2nd ed. as BS-161. 1940. States that 'the livest, the most widespread, and perhaps the most socially significant activity in the field of American biology today is the technology known as wildlife management.'

MCBRIDE, G. McC., and MERLE A. MCBRIDE
1942. 'Highland Guatemala and Its Maya Communities.' *Geog. Rev.* 32:252-68. Includes notes on native agriculture and land use.

MCDOUGALL, W. B.
1941. *Plant Ecology.* Lea & Febiger. 3rd ed., revised. 285 pp. A standard text with emphasis upon relationships, especially symbiotic ones.

MERRIAM, C. HART
1890. *Results of a Biological Survey of the San Francisco Mountain Region and Desert of the Little Colorado, Arizona.* North Amer. Fauna, No. 3. 136 pp. The fundamental study upon which hinged the concept of life zones as Merriam later developed them.

———
1892. 'The Geographic Distribution of Life in North America with Special Reference to the Mammalia.' *Proc. Biol. Soc. Wash.* 7:1-64. Mostly a historical synopsis of life-zone treatments, especially for North America.

———
1894. 'Laws of Temperature Control of the Geographic Distribution of Terrestrial Animals and Plants.' *Nat. Geog. Mag.* 6:229-38. Discussion of the data and method for delineating Merriam's life zones.

———
1898. *Life Zones and Crop Zones of the United States.* U. S. Dept. Agr. Div. Biol. Survey Bul. 10. 79 pp. Lists crop plants for each of Merriam's life zones.

BIBLIOGRAPHY

MERRILL, E. D.

1938. 'Domesticated Plants in Relation to the Diffusion of Culture.' *Bot. Rev.* 4:1-20. Discussion of the importance of American cultivated plants with respect to New World culture.

MILLER, ALDEN H. (EDITOR)

1943. *Joseph Grinnell's Philosophy of Nature: Selected Writings of a Western Naturalist.* Univ. Calif. Press. 237 pp. An outstanding collection of 28 essays on the distribution and environmental relations of birds and mammals of the western United States.

MILLER, J. PAUL, and BURWELL B. POWELL

1942. *Game and Wild-fur Production and Utilization on Agricultural Land.* U. S. Dept. Agr. Cir. 636. 58 pp. A study of the economic aspects of farm game production, with the conclusion that wildlife management is profitable on agricultural land only when it is integrated with other kinds of land management.

MONSON, GALE

1941. 'The Effect of Revegetation on the Small Bird Population in Arizona.' *Jour. Wildl. Mangt.* 5:395-7. Shows birds to be about twice as numerous on properly managed as on overgrazed rangeland.

——

1943. 'Food Habits of the Banner-tailed Kangaroo Rat in Arizona.' *Jour. Wildl. Mangt.* 7:98-102. Demonstrates that this mammal is not common where perennial grasses are dominant, for the latter do not provide as abundant and unfailing crop of seeds as do the annual grasses and weeds.

MONSON, GALE, and WAYNE KESSLER

1940. 'Life-history Notes on the Banner-tailed Kangaroo Rat, Merriam's Kangaroo Rat, and the White-throated Wood Rat in Arizona and New Mexico.' *Jour. Wildl. Mangt.* 4:37-43. Discloses that these rodents eat annual plants for the most part and are most abundant on poor rangeland.

MUNNS, E. N.

1940. *A Selected Bibliography of North American Forestry.* U. S. Dept. Agr. Misc. Publ. 364. 1142 pp. 2 volumes. Comprises 21,413 unannotated references.

MURIE, ADOLPH

1940. *Ecology of the Coyote in the Yellowstone.* U. S. Dept. Int. Fauna Series 4, Cons. Bul. 4. 206 pp. An excellent study of an economically important mammal, especially its relationships with other animals.

MURPHY, ROBERT CUSHMAN

1936. *Oceanic Birds of South America.* Am. Mus. Nat. Hist., 2 vols., 1245 pp.

BIBLIOGRAPHY

NATIONAL RESOURCES COMMITTEE
    1935. *Report on Water Pollution.* Special Advisory Committee on Water Pollution. 82 pp. Processed. Largely a study of pollution with respect to human welfare with list of references.

NATIONAL RESOURCES PLANNING BOARD
    1941. *Land Classification in the United States.* U. S. Gov't Printing Office. 151 pp. Summarizes various systems of classifying land.

NETTING, M. GRAHAM
    1939. 'Reptiles Killed on a "Vermin" Campaign in Mercer County, West Virginia.' *Proc. W. Va. Acad. Sci.* 13:162-6. A critical analysis of the species taken, especially with respect to their economic value.

NICE, M. M.
    1941. 'The Role of Territory in Bird Life.' *Amer. Midl. Nat.* 26:441-87. A summary treatment of the subject by its foremost American student. There is a good bibliography.

NICHOLSON, E. M.
    1927. *How Birds Live: A Brief Account of Bird Life in the Light of Modern Observation.* Williams & Norgate, London. 139 pp. 2nd ed. 1929. 159 pp. Contains a great deal on censusing bird populations.

NORTON, E. A.
    1940. 'Land Classification as an Aid to Soil Conservation Operations.' *Mo. Agr. Expt. Sta. Bul.* 421. pp. 293-304. A discussion of the classification of land according to use capabilities.

OLMSTEAD, CHARLES E.
    1937. 'Vegetation of Certain Sand Plains of Connecticut.' *Bot. Gaz.* 99:209-300. Includes observations on the importance of animals in the establishment of plant communities, as the planting of oaks by squirrels.

OSBORN, BEN
    1940. 'A Photographic Transect for Determining Soil Cover Index of Vegetation.' *Ecology* 21:416-19.

————
    1942. 'Prairie Dogs in Shinnery (Oak Scrub) Savannah.' *Ecology* 23:110-15. An example of animals now living in a community of which they were not originally a part, but which has been made suitable to them by man's use of the land.

PADDOCK, F. B.
    1937. 'Relation of Beekeeping to Wildlife Conservation.' *Amer. Bee Jour.* 77:519-20, 549. Points out value of bees and other insects in pollination and fruit production of plants important to wildlife.

BIBLIOGRAPHY

PAVLYCHENKO, T. K.

1942. 'The Place of Crested Wheatgrass, *Agropyron cristatum* L., in Controlling Perennial Weeds.' *Sci. Agr.* 22:459-60. Emphasizes the value of a knowledge of competition as a control method.

PEARSE, A. S.

1926. *Animal Ecology.* McGraw-Hill. 417 pp. A standard text book, documented.

PETERSON, ROGER T.

1942. 'Life Zones, Biomes, or Life Forms?' *Audubon Mag.* 44:21-30. Points out the importance of life forms of plants and niches in the distribution of birds.

PETRIDES, GEORGE A.

1942. 'Variable Nesting Habits of the Parula Warbler.' *Wilson Bulletin* 54:252-3. In part an argument against the importance of life forms of plants as an influence in the occurrence of animals.

PHILADELPHIA SOCIETY FOR PROMOTING AGRICULTURE

1808-1815. *Memoirs of the Philadelphia Society for Promoting Agriculture.* Johnson and Warner. Vols. 1-3. The older publications of this Society form an interesting and valuable record of progressive thinking among early American agriculturists.

PHILLIPS, JOHN C.

1928. *Wild Birds Introduced or Transplanted in North America.* U. S. Dept. Agr. Tech. Bul. 61. 63 pp. A valuable summary of America's experience with introduced birds.

PIEMEISEL, R. L.

1938. *Changes in Weedy Plant Cover on Cleared Sagebrush Land and their Probable Causes.* U. S. Dept. Agr. Tech. Bul. 654. 44 pp. Indicates that jack rabbits are most numerous on sparsely vegetated land.

PITELKA, FRANK A.

1941. 'Distribution of Birds in Relation to Major Biotic Communities.' *Amer. Midl. Nat.* 25:113-37. A study, with examples, emphasizing the correlation between broad plant communities and the occurrence of birds, and the importance of life forms of plants with respect to bird distribution.

RAUNKIAER, CHRISTEN

1934. *The Life Forms of Plants and Statistical Plant Geography.* Oxford, The Clarendon Press. 632 pp. The fundamental and most exhaustive treatment of the subject. Much too detailed to be of great value to the land-management biologist.

RAUP, HUGH M.

1940. 'Old Field Forests of Southeastern New England.' *Jour. Arn. Arb.* 21:266-73. An excellent example of the value of plants occurring on disturbed areas as indicators of primeval vegetation.

BIBLIOGRAPHY

RENNER, F. G., et al.
1938. A Selected Bibliography on Management of Western Ranges, Livestock, and Wildlife. U. S. Dept. Agr. Misc. Publ. 281. 468 pp. Comprises 8274 unannotated references.

ROBBINS, W. W., A. S. CRAFTS, and R. N. RAYNOR
1942. Weed Control. McGraw-Hill. 543 pp. An excellent text with chapters on principles, competition, life histories, cultural and biological methods of control. Chemical control is treated at length.

SAMPSON, ARTHUR W.
1919. Plant Succession in Relation to Range Management. U. S. Dept. Agr. Bul. 791. 76 pp. One of the earliest and most significant papers on the practical application of the concept of succession to a type of land management.

————
1939. 'Plant Indicators—Concept and Status.' Bot. Rev. 5:155-206. A comprehensive review and bibliography on the subject.

SAUER, CARL O.
1921. 'The Problem of Land Classification.' Annals of the Assoc. of Amer. Geog. 11:3-16. Presents a classification based on the German, bonitierung, system, which contains eight major classes.

SAUNDERS, ARETAS A.
1936. Ecology of the Birds of Quaker Run Valley, Allegany State Park, New York. N. Y. State Museum Handbook 16. 174 pp. A model study of the correlation of the occurrence of birds with vegetation types.

SCHEFFER, PAUL M.
1938. 'The Beaver as an Upstream Engineer.' Soil Conservation 3:178-81.

SCHIMPER, A. F. W.
1903. Plant-geography Upon a Physiological Basis. Oxford, The Clarendon Press. 839 pp. English translation by Fisher, from German ed. of 1898. A classical work upon the distribution of major vegetation types throughout the world.

SCHOENMANN, L. R.
1923. 'Description of Field Methods Followed by the Michigan Land and Economic Survey.' Amer. Assoc. Soil Survey Workers Bul. 4. v. 1, pp. 44-52.

SEARS, PAUL B.
1935. Deserts on the March. 231 pp. Univ. of Okla. Press. A popular but authoritative account of American land use and the value of applying natural principles to human affairs.

BIBLIOGRAPHY

SEARS, PAUL B.

 1939. *Life and Environment: The Interrelations of Living Things.* Teachers College, Columbia Univ. 175 pp. A discussion of human ecology.

——

 1942. 'Forest Sequences in the North Central States.' *Bot. Gaz.* 103:751-61. Illustrates the value of analysis of old pollen deposits in bogs for disclosing the composition of past plant communities.

SHANTZ, HOMER L.

 1911. *The Natural Vegetation as an Indicator of the Capabilities of Land for Crop Production.* U. S. Dept. Agr. Bur. Plant Ind. Bul. 201. 100 pp. One of the earliest scientific papers on the subject.

——

 1917. 'Plant Succession on Abandoned Roads in Eastern Colorado.' *Jour. Ecol.* 5:19-42. This and other studies by the author teach a great deal about vegetation changes in the Great Plains.

——

 1941a. 'The Balance of Nature.' *Colo. Cons. Comments* 4:1-3.

——

 1941b. 'Economic Aspects of Conservation.' *Jour. Forestry* 39:741-7.

SHANTZ, HOMER L., and RAPHAEL ZON

 1924. *Natural Vegetation.* U. S. Dept. Agr. Atlas of Amer. Agr. Natural Vegetation. 29 pp. Description of the original major plant communities of the United States, with large colored map.

SHELFORD, V. E.

 1907. 'Preliminary Note on the Distribution of the Tiger Beetles (*Cicindela*) and Its Relation to Plant Succession.' Marine Biol. Lab. Woods Hole *Biol. Bul.* 14:9-14.

——

 1913. *Animal Communities in Temperate America.* Geog. Soc. of Chicago Bul. 5. Univ. Chicago Press, 362 pp. 2nd ed. 1937. A comprehensive and detailed local study of animal relationships which stands as a major contribution in dynamic animal ecology.

——

 1929. *Laboratory and Field Ecology.* Williams and Wilkins. 608 pp. A standard textbook on methods, devices, instruments, and their use.

——

 1932. 'Life Zones, Modern Ecology, and the Failure of Temperature Summing.' *Wilson Bul.* 44:144-57. Largely a criticism of Merriam's life-zone concept.

BIBLIOGRAPHY

SHELFORD, V. E.
    1942. 'Biological Control of Rodents and Predators.' *Sci. Monthly* 55:331-41. Primarily an appeal for a large area of grassland reserved for study and a more reasonable approach to predator and rodent control work.

SHREVE, F.
    1917. 'A Map of the Vegetation of the United States.' *Geog. Rev.* 3:119-25. Map of original major plant communities with brief descriptions.

SIMPSON, JOHN
    1907. *Game and Game Coverts.* Country Gentlemen's Assoc., London. 84 pp. An early discussion of the relation of wildlife management to agriculture, with recommendations for woodland border plantings and other practices.

SIMPSON, GEORGE GAYLORD, and ANNE ROE SIMPSON
    1939. *Quantitative Zoology.* McGraw-Hill. 414 pp. A useful handbook for the statistically minded biologist.

SMITH, ARTHUR D.
    1940. 'A Discussion of the Application of a Climatological Diagram, the Hythergraph, to the Distribution of Natural Vegetation Types.' *Ecology* 21:184-91. One of the few studies of the application of the climate diagram to the occurrence of plant communities.

SMITH, CHARLES CLINTON
    1940. 'The Effect of Overgrazing and Erosion Upon the Biota of the Mixed-grass Prairie of Oklahoma.' *Ecology* 21:381-397. An exemplary study of changes in plant and animal life as a result of land use.

SNAPP, OLIVER I., and J. R. THOMSON
    1943. *Life History and Habits of the Peachtree Borer in the Southeastern States.* U. S. Dept. Agr. Tech. Bul. 854. 24 pp.

SOPER, J. D.
    1941. 'History, Range, and Home Life of the Northern Bison.' *Ecol. Monographs* 11:347-412.

SPERRY, CHARLES C.
    1941. *Food Habits of the Coyote.* U. S. Dept. Int. Wildl. Res. Bul. 4. 70 pp. Based on an analysis of the contents of 8000 coyote stomachs collected throughout the western United States.

STAPLEDON, R. G.
    1935. *The Land Now and Tomorrow.* Faber & Faber, London. 323 pp. General discussion of land use in England, with reference to recreation, game, tenancy, urban areas, etc.

ST. HILAIRE, I. GEOFFROY
    1854-62. *Histoire naturelle générale des Règnes Organiques, principalement étudiée chez l'homme et les animaux.* V. Masson, Paris. 3 volumes.

BIBLIOGRAPHY

In this work the author used the word ethology to mean the study of living things in relation to environment. St. Hilaire never elaborated later, and Haeckel's term ecology came to express this concept.

STODDARD, HERBERT L.
    1932. *The Bobwhite Quail: Its Habits, Preservation and Increase.* Charles Scribner's Sons. 559 pp. An unparalleled study of a game animal. It includes suggestions for the integration of land use and quail management.

———

    1939. *The Use of Controlled Fire in Southeastern Game Management.* Cooperative Quail Study Assoc., Thomasville, Ga. 21 pp. One of the numerous papers on the subject by one who advocates the use of careful burning as a land-management practice.

STODDART, LAURENCE A., and ARTHUR D. SMITH
    1943. *Range Management.* McGraw-Hill. 547 pp. A standard textbook on one of the major types of land use; well documented.

STORIE, R. EARL
    1933. *An Index for Rating the Agricultural Value of Soils.* Calif. Agr. Expt. Sta. Bul. 556. 44 pp. A description of the method of evaluating the productivity of the soil now known as the Storie Index.

STRONG, LEE A.
    1938. 'Insect and Pest Control in Relation to Wildlife.' *Trans. Third North Amer. Wildlife Conf.* 543-7. Contends that the control of insects and plant diseases is a form of wildlife management.

STUDHOLME, ALLAN T.
    1943. 'Some Census Methods for Game.' *Pennsylvania Game News* 13:10-11, 26-7, 31.

STURDY, D.
    1939. 'Leguminous Crops in Native Agricultural Practice.' *East Afr. Agr. Jour.* 5:31-3. A description of crop culture in Tanganyika.

SWEETMAN, HARVEY L.
    1936. *The Biological Control of Insects: With a Chapter on Weed Control.* Comstock Publ. Co. 461 pp.

SWINGLE, H. S., and E. V. SMITH
    1942. *Management of Farm Fish Ponds.* Ala. Agr. Expt. Sta. Bul. 254. 23 pp. One of the many publications by the authors who, by their work on fishpond management, have demonstrated the practical application of biological principles to the productive use of land.

SYMPOSIUM
    1935. 'Erosion Prevention Capacity of Plant Cover.' *Iowa State Col. Jour. Sci.* 9:323-407. Nine papers by various authors.

BIBLIOGRAPHY

SYMPOSIUM
    1940. 'Symposium on the Relation of Ecology to Human Welfare—The Human Situation.' *Ecol. Monographs* 10:307-72. Six papers by various authors.

TALBOT, M. W., H. H. BISWELL, and A. L. HORMAY
    1939. 'Fluctuations in the Annual Vegetation of California.' *Ecology* 20:394-402.

TANSLEY, A. G.
    1923. *Practical Plant Ecology: A Guide for Beginners in Field Study of Plant Communities.* Dodd, Mead & Co. 228 pp.

————
    1939. 'British Ecology During the Past Quarter Century: The Plant Community and the Ecosystem.' *Jour. Ecol.* 27:513-30. A summary of significant English concepts.

TANSLEY, A. G., and T. F. CHIPP
    1926. *Aims and Methods in the Study of Vegetation.* The British Empire Vegetation Committee. 383 pp. Deals largely with methods, devices, instruments, and their use.

TAYLOR, W. P.
    1930. 'Methods of Determining Rodent Pressure.' *Ecology* 11:523-42. A summary of census methods for small mammals on southwestern rangelands.

THORNTHWAITE, C. WARREN
    1931. 'The Climates of North America According to a New Classification.' *Geog. Rev.* 21:633-55. Description of the author's classification, which is coming to be used rather widely.

————
    1942. *Climatology in the Service of Agriculture.* U. S. Dept. Agr. SCS-MP-25. 9 pp. Mimeo. Emphasizes the importance of a knowledge of microclimates.

TOUMEY, JAMES W.
    1928. *Foundation of Silviculture upon an Ecological Basis.* John Wiley & Sons. 438 pp. 2nd ed., 1937 revised by Clarence F. Korstian, 456 pp. Discussion of environmental factors, classification, and succession of forest communities.

TULL, JETHRO
    1733. *The Horse-hoing Husbandry: or, An Essay on the Principles of Tillage and Vegetation.* Printed for the Author. London. 200 pp. A book which had considerable influence in encouraging cultivation by use of the plow as opposed to hand cultivation.

BIBLIOGRAPHY

TWOMEY, ARTHUR C.

1936. 'Climographic Studies of Certain Introduced and Migratory Birds.' *Ecology* 17:122-32.

U. S. DEPARTMENT OF AGRICULTURE

1938. *Soils and Men.* Yearbook of Agriculture. Gov't Printing Office. 1232 pp. Includes descriptions, with folded, colored map, of the soils of the United States.

———

1941. *Climate and Man.* Yearbook of Agriculture. Gov't Printing Office. 1248 pp. Includes statistical data on the climates of the States of the United States and possessions.

U. S. SENATE

1933. *A National Plan for American Forestry.* U. S. Senate Document 12. 2 volumes. 1677 pp. A comprehensive discussion of forest resources and management.

———

1936. *The Western Range.* U. S. Senate Document 199. 620 pp. A comprehensive discussion of the range resources of the United States and their management, with bibliography.

UVAROV, B. P.

1931. 'Insects and Climate.' *Trans. Ent. Soc. London* 79: 1-247.

VANDERSAL, WILLIAM R.

1938. *Native Woody Plants of the United States: Their Erosion-control and Wildlife Values.* U. S. Dept. Agr. Misc. Publ. 303. 362 pp. Includes a revised version of Mulford's map of plant growth regions, and maps showing correlation of those regions with soils and climates, respectively, of the United States.

———

1937. 'The Dependence of Soils on Animal Life.' *Trans. Second North Amer. Wildlife Conf.* 458-67. Summary discussion of the effects of animals upon soils, with bibliography.

———

1940. 'Environmental Improvement for Valuable Non-game Animals.' *Trans. Fifth North Amer. Wildlife Conf.* 200-202. A short but pointed article contending that increase of many kinds of birds and mammals will depend primarily upon the kinds of general land use practices adopted.

———

1942. *Ornamental American Shrubs.* Oxford Univ. Press. 288 pp. A readable book on attractive shrubs native to the United States, and the advantages of encouraging non-exotic species.

[258]

BIBLIOGRAPHY

VANDERSAL, WILLIAM R.

1943. *The American Land: Its History and Its Uses.* Oxford Univ. Press. 215 pp. Popular description of crop plants grown in the United States, with notes on land use.

VAN HISE, CHARLES R., and LOOMIS HAVEMEYER

1937. *Conservation of Our Natural Resources.* Macmillan. 551 pp. Based on *Conservation of Natural Resources in the United States*, by Van Hise, 1910. This book remains one of the best of the many that have appeared on conservation.

VEATCH, J. O.

1933. *Agricultural Land Classification and Land Types of Michigan.* Mich. Agr. Expt. Sta. Spec. Bul. 231. 51 pp.

WARMING, E.

1909. *Oecology of Plants: An Introduction to the Study of Plant Communities.* Oxford, The Clarendon Press. 422 pp. 2nd ed., 1925. This is the English translation by Groom and Balfour from *Plantesamfund*, Danish ed., 1895. German editions appeared in 1896 and 1902. The earliest comprehensive work on the dynamic aspects of plant ecology.

WASHINGTON STATE PLANNING COUNCIL, LAND USE COMMITTEE

1934. *Report of the Land Use Committee, Washington State Forestry Conference.* 7 parts. Mimeo. Prepared by D. S. Jeffers, et al.

WEAVER, HAROLD

1943. 'Fire as an Ecological and Silvicultural Factor in the Ponderosa-pine Region of the Pacific Slope.' *Jour. Forestry* 41:7-14. Contends that ponderosa pine forests of California, Oregon, Washington, and Idaho owe their existence to fire and that areas protected from burning 30 to 40 years show reproduction of Douglas Fir, White Fir, and Incense Cedar.

WEAVER, J. E., and F. W. ALBERTSON

1939. 'Major Changes in Grassland as a Result of Continued Drought.' *Bot. Gaz.* 100:576-91. Shows that continued drought can alter even the dominant perennial species of a community.

WEAVER, JOHN E., and FREDERICK E. CLEMENTS

1929. *Plant Ecology.* McGraw-Hill. 520 pp. 2nd ed., 1938, 601 pp. A standard textbook, well documented.

WEBB, WALTER PRESCOTT

1936. *The Great Plains.* Houghton Mifflin. 525 pp. Contains an interesting account of the history of fencing in the West and discussion of the famous barb-wire controversy.

WEESE, A. O.

1939. 'The Effect of Overgrazing on Insect Population.' *Proc. Okla. Acad. Sci.* 19:95-9. Compares the insects of heavily grazed with essentially undisturbed bluestem range.

[259]

BIBLIOGRAPHY

WEHRWEIN, GEORGE S.

1942. 'The Rural-Urban Fringe.' *Econ. Geog.* 18:217-28 Shows that the ecotone has economic as well as biologic significance.

WEISS, PAUL

1942. 'The Training of Biologists.' *Science* 95:32-4. Points out the need for a sense of social responsibility on the part of biologists.

WELLS, H. G.

1942. 'Is This the Last War?' *The Sunday Star*, Washington, D.C., 31 May. Emphasizes the need for human ecologists or 'ministers of foresight' to help solve the problems of mankind.

WENT, F. W.

1942. 'The Dependence of Certain Annual Plants on Shrubs in Southern California Deserts.' *Bul. Torrey Bot. Club* 69:100-114.

WHEELER, C. M., J. R. DOUGLAS, and F. C. EVANS

1941. 'The Role of the Burrowing Owl and the Sticktight Flea in the Spread of Plague.' *Science* 94:560-61. Reports the owl to harbor plague-carrying fleas.

WHITAKER, H. L.

1939. 'Fox Squirrel Utilization of Osage Orange in Kansas.' *Jour. Wildl. Mangt.* 3:117. Suggests that the westernmost extension of the fox squirrel coincides with the extension of Osage Orange, a food of the squirrel.

WHITFIELD, C. J., and C. L. FLY

1939. 'Vegetational Changes as a Result of Furrowing on Pasture and Range Lands.' *Jour. Amer. Soc. Agron.* 31:413-17.

WHITSON, A. R.

1935. 'Land Classification in Wisconsin.' *Amer. Soil Survey Assoc. Bul.* 16:39-41.

WHYTE, R. O.

1940. 'The Control of Weeds.' *Imp. Bur. of Pasture and Forage Crops, Herb. Publ. Series Bul.* 27. 169 pp.

WIGHT, HOWARD M.

1933. *Suggestions for Pheasant Management in Southern Michigan.* Mich. Dept. Cons. 25 pp. Contains an early reference to the possibility of integrating wildlife habitat improvements and erosion control.

―――

1938. *Field and Laboratory Technic in Wildlife Management.* Univ. of Mich. Press. 107 pp. Deals largely with methods, particularly the preparation of specimens.

WILBUR, D. A., R. F. FRITZ, and R. H. PAINTER

1942. 'Grasshopper Problems Associated with Strip Cropping in Western Kansas.' *Jour. Amer. Soc. Agron.* 34:16-29.

BIBLIOGRAPHY

WILLIAMSON, LYMAN O., and D. JOHN O'DONNELL
    1941. 'Muskie Food.' *Wis. Cons. Bul.* 6:14-18.

WINSHIP, G. P.
    1896. 'The Coronado Expedition (1540-42).' *Fourteenth Annual Report of the Bureau of American Ethnology.* Part I. pp. 329-67. Gov't Printing Office. Includes descriptions of virgin conditions in the Great Plains and Southwest.

WODEHOUSE, R. P.
    1940. 'Hold That Sneeze!' *Rotarian* 57:43-44. An authority on pollen states that good land use, by reducing weeds, may reduce hay fever.

WYGANT, N. D.
    1941. 'An Infestation of the Pandora Moth, *Coloradia pandora* Blake, in Lodgepole Pine in Colorado.' *Jour. Econ. Entomology* 34:697-702.

ZIERER, CLIFFORD M.
    1941. 'Brisbane—River Metropolis of Queensland.' *Econ. Geog.* 17:325-44.

ZINSSER, HANS
    1937. *Rats, Lice, and History.* N. Y. Blue Ribbon Books. 313 pp. Popular life history of typhus fever and its effect upon the course of human events.

SCIENTIFIC NAMES

THE following list of vernacular names with their scientific equivalents will serve to identify the plants and animals mentioned in the text. The list does not include generally understood generic terms, like oak and rat, or familiar names, such as carp and lady beetle, for which identifications are provided in standard unabridged dictionaries.

Antelope: *Antilocapra americana*
Aphid, woolly apple: *Eriosoma lanigerum*
Armadillo: *Dasypus novemcinctus*
Army worm: *Cirphis unipuncta*
Arrow-wood: *Viburnum dentatum*
Ash, single-leaf: *Fraxinus anomala*
  white: *Fraxinus americana*
Aspen: *Populus aurea* (western); *P. tremuloides* (eastern)
Auk, great: *Plautus impennis*

Badger: *Taxidea taxus*
Bass: *Huro salmoides*
Basswood: *Tilia heterophylla*
Beech: *Fagus grandifolia*
Beet leaf hopper: *Eutettix tenellus*
Beetle, Colorado potato: *Leptinotarsa decemlineata*
  Japanese: *Popillia japonica*
  white-fringed: *Pantomorus leucoloma* and related species
Bindweed: *Convolvulus arvensis*
Birch, black: *Betula lenta*
  gray: *Betula populifolia*
  paper: *Betula papyrifera*
  red: *Betula nigra*
  white: *Betula papyrifera* or *B. populifolia*
  yellow: *Betula lutea*
Blacksnake, pilot: *Elaphe obsoleta*
Blueberry: *Vaccinium*

Bluegrass: *Poa pratensis*
Bluestem, little: *Andropogon scoparius*
Boll worm, pink: *Pectinophora gossypiella*
Bream: *Lepomis*
  bluegill: *Lepomis macrochirus*
Brome, mountain: *Bromus carinatus*
'Brown spot': *Septoria acicola*
Buffalo: *Bison bison*
Bug, assassin: *Reduviidae*
  chinch: *Blissus leucopterus*
  damsel: *Nabis* or related genera
Bugloss: *Echium vulgare*
Bulrush: *Scirpus*
Buttonbush: *Cephalanthus occidentalis*

Cactus, giant: *Cereus giganteus*
  prickly-pear: *Opuntia*
Cape weed: *Cryptostemma calendulacea*
Cat, ring-tailed: *Bassariscus astutus*
Catfish: *Ameiurus*
Cat-tail: *Typha latifolia*
Cedar, incense: *Libocedrus decurrens*
  red: *Juniperus virginiana*
  western red: *Thuja plicata*
Cherry, black: *Prunus serotina*
  pin: *Prunus pennsylvanica*
Chestnut: *Castanea dentata*

SCIENTIFIC NAMES

Chestnut blight: *Endothia parasitica*
Chipmunk: *Eutamias* (western); *Tamias* (eastern)
Cisco: *Leucichthys artedi* or related species
Cliff rose: *Cowania stansburiana*
Codling moth: *Carpocapsa pomonella*
Corn borer, European: *Pyrausta nubilalis*
Corn-ear worm: *Heliothis obsoleta*
Cotton boll weevil: *Anthonomus grandis*
Cotton boll worm: *Heliothis obsoleta*
Cotton flea hopper: *Psallus seriatus*

Deer, mule: *Odocoileus hemionus*
Virginia: *Odocoileus virginianus*
Dropseed: *Sporobolus*

Elk: *Cervus canadensis*

Fanweed: *Thlaspi arvense*
Fir, balsam: *Abies balsamea*
Douglas: *Pseudotsuga taxifolia*
white: *Abies concolor*
Fireweed: *Epilobium* and *Erechtites*
Flea, sticktight: *Echidnophaga gallinacea*
Fly-weevil: *Sitotroga cerealella*
Fox, arctic: *Alopex lagopus*
grey: *Urocyon cinereoargenteus*

Gall rust, Woodgate: *Cronartium*
Goose, blue: *Chen caerulescens*
snow: *Chen hyperborea*
Grama: *Bouteloua*
Grass, cheat: *Bromus tectorum*
orchard: *Dactylis glomerata*
poverty: *Danthonia spicata*
quack: *Agropyron repens*
salt: *Distichlis stricta*
Greasewood: *Sarcobatus vermiculatus*
Ground squirrel: *Citellus*
Harris: *Citellus harrisii*
Groundhog: *Marmota monax*
Grouse: *Bonasa* or closely related genera
black: *Lyrurus tetrix*

Grouse (Cont.)
pinnated: *Tympanuchus cupido americanus*
ruffed: *Bonasa umbellus*
Gum, black: *Nyssa sylvatica*
red: *Liquidambar styraciflua*
Gypsy moth: *Porthetria dispar*

Hackmatack: *Larix laricina*
Hare: *Lepus bairdi*
snowshoe: *Lepus americanus*
Hawk, Cooper's: *Accipiter cooperi*
marsh: *Circus hudsonius*
red-tailed: *Buteo borealis*
sparrow: *Falco sparverius*
Hazelnut, California: *Corylus rostrata* var. *californica*
Heath hen: *Tympanuchus cupido cupido*
Hemlock: *Tsuga canadensis*
western: *Tsuga heterophylla*
Hessian fly: *Phytophaga destructor*
Hobblebush: *Viburnum alnifolium*
Horned lark: *Otocoris alpestris*

Ironweed: *Vernonia*

Jackrabbit, Allen's: *Lepus alleni*
blacktailed: *Lepus californicus*

Kangaroo rat, banner-tailed: *Dipodomys spectabilis*
Kingfisher: *Megaceryle alcyon*
Kudzu: *Pueraria thunbergiana*

Larch, American: *Larix laricina*
Locust, black: *Robinia pseudoacacia*

Maple, hard: *Acer saccharum*
red: *Acer rubrum*
sugar: *Acer saccharum*
Marmot: *Marmota flaviventris*
Marten, pine: *Martes americana*
Mealy bug: *Pseudococcus citri* and related species
Mesquite: *Prosopis chilensis*
Mormon cricket: *Anaprus simplex*
Morning glory: *Convolvulus arvensis*
Mouse, climbing deer: *Peromyscus*
field: *Microtus*
meadow: *Microtus*

Mouse (Cont.)
pine: *Pitymys pinetorum*
white-footed: *Peromyscus*
Mynah: *Acridotheres tristis*

Oak, black: *Quercus velutina*
red: *Quercus borealis*
scarlet: *Quercus coccinea*
Spanish: *Quercus coccinea*
white: *Quercus alba*
Owl, burrowing: *Speotyto cunicularia*
great-horned: *Bubo virginianus*

Palm, royal: *Roystonea regia*
Palmetto: *Sabal minor*
Partridge, Chukar: *Alectoris chukar*
European: *Perdix perdix*
Hungarian: *Perdix perdix*
Pea-bug: *Mylabris pisorum*
Peachtree borer: *Aegeria exitiosa*
Perch, yellow: *Perca flavescens*
Pine, jack: *Pinus banksiana*
Jeffrey: *Pinus jeffreyi*
loblolly: *Pinus taeda*
lodgepole: *Pinus murrayana*
longleaf: *Pinus palustris*
Monterey: *Pinus radiata*
Norway: *Pinus resinosa*
pitch: *Pinus rigida*
ponderosa: *Pinus ponderosa*
Scotch: *Pinus sylvestris*
scrub: *Pinus virginiana*
short-leaf: *Pinus echinata*
Virginia: *Pinus virginiana*
white: *Pinus strobus*
yellow: *Pinus ponderosa*
Pinyon: *Pinus edulis*
Plover, golden: *Pluvialis dominica*
Pocket gopher, California: *Thomomys bottae*
Prairie chicken: *Tympanuchus cupido*
Prairie-dog, black-tailed: *Cynomys ludovicianus*
Gunnison: *Cynomys gunnisoni*
Pyrethrum: *Chrysanthemum cinerariaefolium*

Quail, bobwhite: *Colinus virginianus*
California: *Lophortyx californica*
Gambel: *Lophortyx gambeli*
scaled: *Callipepla squamata*

Rabbit, European: *Oryctolagus cuniculus*
Ragweed: *Ambrosia artemisiifolia* or *A. trifida*
Raven, white-necked: *Corvus cryptoleucus*
Redstart: *Setophaga ruticilla*
Redtop: *Agrostis alba*
Road-runner: *Geococcyx californianus*

Sage hen: *Centrocercus urophasianus*
Sagebrush: *Artemisia tridentata*
little: *Artemisia nova*
Sauger: *Stizostedion canadense*
Scale, black: *Saissetia oleae*
cottony cushion: *Icerya purchasi*
San José: *Aspidiotus perniciosus*
Shadscale: *Atriplex canescens* or related species
Sheep sorrel: *Rumex acetosella*
Sheepshead: *Aplodinotus grunniens*
Shrew, least: *Microsorex hoyi*
short-tailed: *Blarina brevicauda*
Shrike: *Lanius ludovicianus*
Skunk, hognosed: *Conepatus mesoleucus*
spotted: *Spilogale arizonae*
Spruce, red: *Picea rubra*
Squirrel, fox: *Sciurus niger*
gray: *Sciurus carolinensis*
red: *Sciurus hudsonicus*
Sycamore: *Platanus wrightii* (western); *P. occidentalis* (eastern)

Thistle, Canada: *Cirsium arvense*
Russian: *Salsola pestifer*
sow: *Sonchus arvensis*
Toad flax: *Linaria vulgaris*
Tulip tree: *Liriodendron tulipifera*
Tupelo: *Nyssa aquatica*

Vervain: *Verbena hastata*

Walnut, black: *Juglans nigra*
Warbler, Parula: *Compsothlypis americana*
Water caltrop: *Trapa natans*
Water hyacinth: *Heteranthera crassipes*

SCIENTIFIC NAMES

Water ouzel: *Cinclus mexicanus*

Wax myrtle: *Myrica cerifera*

Weevil, alfalfa: *Phytonomus posticus*

  southern cowpea: *Mylabris chinensis*

Wheat-fly: *Sitotroga cerealella*

Wheatgrass, crested: *Agropyron cristatum*

  western: *Agropyron smithii*

White grub: *Phyllophaga* or *Lachnosterna*

White pine blister rust: *Cronartium ribicola*

Whitefish: *Coregonus clupeaformis*

Wild oats: *Avena fatua*

Wood rat, Attwater: *Neotoma floridana attwateri*

Yellow rocket: *Barbarea vulgaris*

INDEX

INDEX

Conservation, 170, 229, 230, 231
Conservation practices, soil, 111, 182, Pls. 9, 10
Contour cultivation, 6, 101, 113, 115, Pls. 9, 13, 27
Contour furrows, 159, 160, Pls. 10, 12
Contour ridges, 6, 159
Control, biological, 198, 200, 201, 212
  insect, 106, 113, 114, 115, 116, 217
  noxious plant, 217
  plant disease, 219
  pondweed, 190
  predator, 62, 148, 151, 162, 214
  rodent, 52, 104, 114, 148, 152, 156, 209, 215
  weed, 52, 104, 112, 116, 146, 209, 217
Control criteria, 213
Control methods, 211
  biological, 212
  chemical, 212
  cultural, 212
  mechanical, 212
Control of organisms, 208
Cormorant, 44
Corn, 6, 21, 174, 186, 202
Corn borer, European, 21, 54, 85, 116
Corn-ear worm, 116
Cosmos, 20
Cotton, 202
Cotton boll weevil, 54, 116, 195
Cotton boll worm, 116
Cotton flea hopper, 116
Cottontail, 58, 61, 81, 132, 166, 169
Coyote, 27, 53, 58, 131, 148, 149, 152, 213
  food of, 149, 150
Cranberry, 47, 173
Crane, 185
Crop rotation, 5, 106, 113, 190
Crop yields, 111
Cross timbers, 24
Crotalaria, 5
Croton, 116
Crow, 154
Cruising radius, 59, 60, 162

Crustacean, 188, 189
Culture and land, 225, Pl. 1
Curly top, of sugar beet, 157
Cycles, 66, 152

Dahlia, 20
Dauerwald, 128
Deer, 51, 57, 79, 80, 81, 130, 132, 133, 135, 165, 221, Pl. 19
  mule, 60, 61, 131
  Virginia, 61
Deer drive, 80
Dinosaur, 12
Diptera, 158
Distribution, 17, 59
Ditchbank, irrigation and drainage, 173, 175, 186, 217
Dog, 6, 227
Dogwood, 43, 48, Pl. 2
Drainage, 186, 191
Dropseed, Pl. 14
Duckbill platypus, 18
Ducks, 82, 150, 185, Pl. 30
Dust Bowl, 39, 69, 104, 143

Eagle, bald, 66
Earthworms, 25, 108
Ecology, 8, 11, 72, 111, 228, 230
Ecotone, 11, 38, 124, 141, 167, 168
Edaphic climax, 24
Edge, 38, 141, 167, 168, 186, Pl. 25
Eland, 58
Elk, 60, 79, 80, 131, 133, 149
Elm, 168, Pl. 4
Elodea, 197
Eltonian pyramid, 58
Enclosure, 77
Ephedra, 51
Epilobium, 128
Epiphytes, 29
Erechtites, 128
Erosion, 42, 51, 78, 106, 107, 147, 185, 223, Pls. 12, 20, 27, 28, 30
Erosion silt, in streams, 181, 182, 183
Ethology, 9
Eucalyptus, Pl. 18
Euphorbia esula, 218
Exclosure, 77, Pl. 8
Exotics, 19, 161, 164, 194

[268]